Design & Operation of Wind Farm Support Vessels 2019

London, United Kingdom
27 February 2019

ISBN: 978-1-5108-9543-0

Printed from e-media with permission by:

Curran Associates, Inc.
57 Morehouse Lane
Red Hook, NY 12571

Some format issues inherent in the e-media version may also appear in this print version.

The Royal Institution of Naval Architects
8-9 Northumberland Street
London, WC2N 5DA
United Kingdom

Phone: 020 7235 4622
Fax: 020 7259 5912

publications@rina.org.uk

Additional copies of this publication are available from:

Curran Associates, Inc.
57 Morehouse Lane
Red Hook, NY 12571 USA
Phone: 845-758-0400
Fax: 845-758-2633
Email: curran@proceedings.com
Web: www.proceedings.com

TABLE OF CONTENTS

TECHNICAL & REGULATORY DEVELOPMENTS TO SUPPORT INNOVATIVE OFFSHORE WIND VESSEL

E. Peño, Bureau Veritas, France

SUMMARY

Over the last few years offshore wind industry has become a key contributor to the transformation of the world energy sources. This transformation is driven not only by the increased demand of energy coming from emerging countries but also by the increased awareness about climate change.

To keep on developing this industry, offshore wind farms are now moving further from shore and into deeper water. In addition, to fight back against other energy sources, an outstanding effort is requested to the whole supply chain, including vessels serving this industry, to reduce costs. This paper will explain how the development of new technologies certainly helps to solve the complex equation that the offshore wind industry is facing today. In addition, this paper will describe how classification societies support these developments by providing a constructive regulatory scope to ensure that this innovative vessels can meet this challenge without compromising on safety and while reducing their environmental impact.

NOMENCLATURE

CO2	Carbon dioxide
SOx	Sulphur oxide
NOx	Nitrogen oxide
LNG	Liquefied Natural Gas
O&M	Operating and Maintenance

1. INTRODUCTION

The offshore wind industry is a fast-growing, fast-evolving market with a promising outlook. Of course ships engaged in providing support to this industry will need to keep up with this constant evolution to be ahead of the game in an increasingly competitive market.

In addition, offshore wind industry has the dubious honor of having carried out an impressive reduction of costs in a remarkable short period of time, this cost reduction is achieved by putting a significant pressure on the full supply chain, including ships supporting this industry.

Finding the right balance between the real industry needs and the ship value can prove quite a conundrum for ship owners and ship operators.

In this context Bureau Veritas regards innovation as the fastest way to achieve the required performance while remaining at a sustainable costs. As a consequence, the role of a classification society needs to evolve as well; by continuously updating rules and guidelines, Bureau Veritas helps to ensure that new designs and technologies can be safely implemented on board BV classed vessels.

2. FACING THE CHALLENGE OF THE LCOE

The Levelised Cost Of Energy (LCOE) makes the offshore wind industry go round overall with the zero-subsidy bids debate getting momentum. Today it's paramount to find ways to reduce capital and operational costs while remaining sustainable and technology innovations can help to meet the challenge. Just to mention some of these innovations:

2.1 USE OF BATTERIES

There is today a growing push to make energy storage and offshore wind work in tandem.

The use of batteries will have a positive impact on the produced energy value by controlling the variability of the produced energy, balancing the supply and the demand to avoid price variations that may affect consumers.

On the other hand, in a cost reductions driven industry it may seem paradoxical to add an extra expense to an offshore wind farm development, but ensuring stability and reliability in the energy supply by combining storage and wind energy seems to create a positive synergy in terms of cost reduction.

Additionally having a storage capability allows the integration of other sources of renewable energy such a solar or tidal power.

2.2 INNOVATIVE TURBINE INSTALLATION STRATEGY

Installation at sea is expensive and complex. In addition, turbine installation is highly dependent on weather conditions and consequently downtime caused by weather plays a significant role in the final installation costs.

Defining new pre-assembly strategies that maximize the number of components assembled at port can definitively help to overcome these difficulties and reduce the installation costs. Furthermore, pre-assembly at port can be carried out in parallel to installation at sea which will reduce even more the installation time and therefore installation costs.

2.3 HIGHER RATED CAPACITY OF CABLES

As just mentioned for the installation strategies, cables also take a big part of the capital expenditure in the development of an offshore wind farm. Improving their efficiency in terms of power transmission will have an impact on the sizing of the export system and consequently on its cost, for example by reducing the number of cables even by reducing the need of substations.

Again, the challenge is to achieve the right balance between the cables advanced technology cost and the cost reduction due to the increase of the cable effectiveness.

2.4 STANDARDIZATION

Standardization benefits have been well proven in other industries so why not for this one? Moving to standardization, including foundations, will certainly simplify the training of technicians and also it will facilitate the hold spares just to mention some of the benefits.

In particular regarding foundations, today monopiles foundations are the most commonly used foundation and also monopiles show the highest degree of standardization, chicken or egg paradox?

2.5 TURBINE SIZE

Definitively all the recent improvements to optimize the turbine design (rotors, blades, nacelle) can only help to reduce the LCOE.

However there is a general agreement to say that the most important driver of cost reduction is the turbine size.

In fact, a larger turbine will not increase significantly the efficiency however the fact that fewer turbines need to be installed to deliver the same power will have a positive impact on the capital expenditure:

- Fewer foundations structures
- Simplified power export system

As well as the operating costs:

- Less maintenance
- Less spare parts

And now we are all witnessing a very interesting race; the size of the turbines grows by the day, a new rated power record is broken soon after its announcement. The possibility of having in the future 15 MW even 20 MW turbines is making the headlines today.
Obviously this race to getting bigger will impact the ships supporting this industry as described in later in this paper

3. GO DEEPER, GO FURTHER

As the offshore wind industry develops, wind farms need to move further offshore and explore new regions which may imply in some cases also moving into deeper waters. Besides, there are some other reasons why the wind farms may need to move further from shore:

- Take advantage of better weather conditions
- Avoid poor seabed conditions
- Deep coast area
- Avoid noise and visual pollution
- Avoid interfering with other industries such as tourism or fishing

Of course, going deeper and further imply a greater technical challenge and consequently even more difficult constraints to reduce cost.

In this context, the option of floating offshore turbines is gearing up. The promising outlook of this option is on the table, whether this technology will move from concept or prototype project (Scotland, France...) to full scale development will depend on the feasibility of overcoming the cost barrier.

4. IMPACT ON SERVICE LOGISTICS

4.1 INSTALLATION VESSELS

As a summary of this paper so far; offshore wind farms need to move further from shore and need to reduce costs by mainly increasing the size of the turbines.

Definitively all the different kinds of ships servicing this industry will be affected since they will need to adapt to the same constrains. But if we have to nominate a winner, installation vessels would have the dubious honour of being the most challenged type.

Indeed, installation is a costly operation even more if the travel distance is increased and of course the bigger the turbine the bigger the installation vessels need to be and the bigger the lifting capacity.

In addition, as we have seen previously in this paper, new installation strategies can help to reduce the installation costs, for example by establishing new pre-assembly strategies at port which on the one hand it will reduce installation at sea time but on the other hand it will increase the need for larger vessels with increased lifting capacity.

And the need of increased lifting capacity brings us to the question of safe lifting operation in floating installation units.

From a safety point of view, lifting is a critical operation. Fortunately this risk has been acknowledged and properly addressed at IMO level by adopting the new amendments

to 2008 IS Code. The main objective of these new amendments is to ensure that the ship can carry out the expected lifting operation while remaining within the stability safety limits even after a sudden loss of hook load when using counter ballast.
These amendments have been adopted by MSC.415(97), entering into force in January 2020. The new requirements are included in Part B of the 2008 IS Code, meaning that they are considered just as a recommendation. However Bureau Veritas has already implemented these requirements for ships built after 2018 in order to ensure that the new constructions can carry out lifting operation in a safely manner.

4.2 TRANSFER OF TECHNICIANS

As explained above moving further away from shore certainly adds complexity to the installation part but moving further away also adds complexity to the transfer of technicians which represents a significant part of the O&M costs.

First of all, when travel time is increased it may not be possible for the technicians to simply go back home after a working day, so it may be necessary to find ways to accommodate people at sea and to transfer them from the accommodation to the turbines as well as to shore.

When technicians remain at sea for longer periods of time, whenever the weather conditions allow a potential window for maintenance, the accommodation and transfer strategy shall:

- Ensure that that technicians are in an safe and suitable working condition, that is, ensure that the accommodation provides an adequate level of comfort in terms of accelerations, noise, vibrations…when assessing the level of comfort, we should always keep in mind that technicians are not seamen so they may have a lower level of tolerance to seasickness

- Optimize the number of transfers during the available operational window. For that it becomes paramount to find the right transfer strategy by a combination of new innovative transfer means (such as daughter crafts and walk-to-work systems) in order to reduce the travel time and increase the redundancy.

To meet this complex challenge a new generation of highly innovative vessel have seen the light of day.

- Efficient Service Operating Vessel (SOV) combining high level of operability, low operating cost and advanced transfer logistic solution (gangways, daughter crafts...)

- Crew transfer vessel (CTV) becoming more polyvalent and providing higher personnel carriage capacity.

Recently, there is a new trend of "hybrid designs", halfway between Service Operation Vessels and Crew Transfer Vessels, where technicians can be accommodated and transferred directly to turbines without a gangway. This design can be an interesting alternative solution for certain wind farms subject to a certain condition (distance from shore, number of turbines…).

However, as long as international regulation is concerned, the one-million-dollar question is: who are these technicians when it comes to apply international regulations such as SOLAS?
IMO tools in general and SOLAS in particular make reference to "passengers" and to "crew" and even to "special personnel" (those who are not crew nor passenger but who carry out a special work on board in connection to the main activity of the vessel). However the technicians don't fall into any of these categories. How shall we consider them?

One may think that this issue is old news, the question of applicable regulations for the transfer of more than 12 personnel to and from Oil&Gas offshore facilities, such as FPSO, has been going on for while however the rapid development of the offshore wind industry has pushed the subject again to the front line.

In response IMO has established a working group to develop a new code applicable for ships engaged in the carriage of technicians. Because this new Code will be introduced by a new chapter in SOLAS, it will be mandatory, the estimated date of entering into force is 2024. In the meantime, IMO has issued an interim recommendation, the MSC.418(97) where there are some interesting points. The first one is that a name has been giving to this new personnel category: "industrial personnel". In addition, the recommendation points to existing regulations which are considered as providing an acceptable level of safety. As far as conventional ship is concerned (that is ship which are not high speed crafts) this interim solution points to 2008 SPS Code, however a grey area remains for high speed craft…

In addition, with an advanced experience in the sector North European countries has developed their own rules applicable for ship flying their flag.

5. IMPACT ON SHIP'S DESIGN & EQUIPMENT

This paper has explained so far how these game changers (cost reduction, location, turbine size…) impact on the type and size of vessels but there is more than that. When it comes to reduce costs and in a context where the International Regulations become more and more severe with respect to the environmental protection, the way that the ship is designed and equipped becomes crucial for a ship intended to operate in this highly competitive market.

5.1 ENVIRONMENTALLY FRIENDLY AND EFFICIENT

Even if originally the first role of IMO concerned mainly maritime safety, additionally the Organization has assumed over the years the responsibility for providing measures against marine pollution, originally focus on water pollution by oil. Progressively their commitment, directed by the Marine Environment Protection Committee (MEPC), has been enlarged to include additional measures against other types of water pollution (chemical, sewage…) as well as measures against air pollution and emission from ships.

It's not the purpose of this paper to go through all these emission regulations however the output of the MEPC 72 merits a particular attention. There is a general agreement to consider the output of this committee (which took place some months ago) as historical. During this committee IMO decided to align with Paris change agreement resulting in the adoption of a very ambitious strategy to reduce drastically the emissions of Green House Gases and CO2, by 2050.

In other words, as severe as we may find the regulation today, even more severe emission regulations are to be expected in the future.

In addition, apart from the obvious benefits for the planet, finding ways to reduce emission (for example by enhancing energy efficiency) may have also a beneficial side effect in terms of operating costs reduction.

Again, it is not the purpose of this paper but mainly there are 3 ways to comply with the current regulation: choose the right fuel, clean the exhaust gas or optimise performance over operational mode.

Regarding the choice of fuel, of course, using sulphur compliant fuel will solve the problem of the SOx emissions. Regarding NOx emission, some other alternative fuels may be considered as the LNG. The benefits of LNG as fuel in terms of NOx and SOx emission are undeniable, there are safety risks when using LNG that have to be properly addressed (identification of hazardous area, tanks storage, gas handling equipment…). These risks have been addressed and efficient mitigation measures have been agreed at IMO level with the adoption of the IGF Code.

But in addition to the "traditional" solutions (alternative/compliant fuel or clean the exhaust gases) , a solution could be also to find a way to optimize performance over operational mode, in other words, find a way to ensure that the engines will be always working at their most efficient point and that for every possible load situation.

Last generation of power electronic opens the door to this possibility. Today there is in the market a new generation of compact and efficient drivers and converters that allows converting AC to DC and DC into variable AC. Choosing the right place within the network architecture, the AC switchboard can be replaced by DC grids. As a consequence, generators are no longer locked at fixed rpm in order to keep the frequency of the network, the rpm can be optimized depending on the demand.

The first obvious advantage is the reduction of fuel consumption and consequently the reduction of emissions. However there are some other advantages, for example, the improvement of comfort because of the reduction of noise and vibration. At last but not least, an electric network with DC Grid is ready for batteries.

The combination of DC grids and batteries will give the highest level of optimization on board. If the produced energy exceeds the demand, the energy can be stored. If there is a peak of power demand, it can be covered by batteries instead of starting another engine. As a result, the system will be able to operate optimally as far as emission reduction and fuel consumption is concerned.

However, the complexity of battery systems is to be addressed properly mainly regarding the power management system. In addition, there are some safety risks, mostly the thermal run away, that need to be taken thoughtfully into consideration. Bureau Veritas notation Electric-hybrid addresses the complexity of these system defining requirements for the store, the power management system, control and instrumentation as well as tests that must be carried out in order to ensure that ship assigned with this notation will take fully advantage of this technology in safely manner.

5.2 HIGH LEVEL OF OPERABILITY

When it comes to reduce costs and increase productivity the worst enemy is the weather conditions. The amount of downtime caused and the risks introduced by weather have significant impact on the O&M cost of offshore wind turbines.

On the positive side, there are some solutions to enlarge the weather window

5.2 (a) Hull optimization

Using numerical analysis techniques, such as computational fluid dynamics (CFD), to simulate ship's motion and acceleration, the hull form can be optimized to improve seakeeping capabilities and consequently improve the workability.

5.2 (b) Walk-to-work and Dynamic Positioning operability analysis

As mentioned earlier in the paper, in order to increase productivity and efficiency, the motion compensated walk-to-work system have become almost a common feature for Service Operation Vessel (SOV).

More performant systems continue to enter the market with greater capacity, working at a wider range of heights and harsher weather conditions in terms of wave height. However, integration of the motion compensated gangways and the dynamic position system is a major technical challenge.

The dynamic positioning system is responsible of keeping the ship within the operational limits in the horizontal plane, the motion compensated system is responsible of mitigating the sea effects. The degree of accuracy of the DP sensor data shall be properly interfaced with the motion compensation in order to maximize the weather operation window.

5.2 (c) Gangways system

As strange as it may seems, even if the gangway systems are common place today as mentioned in the previous paragraph, there is no international standard or regulation covering this kind of systems.

To fill this gap, Bureau Veritas has developed a guidelines and notation. The objective is to provide guidance for the design, structural assessment, control and testing systems of gangways system.

5.3 CYBER ENABLED

Shipping industry is known for being a conservative industry quite reluctant to change but the time has come to undergone its own digital transformation as other industries, as aviation, have already done in the past. This transformation raises a certain number of questions.

First of all, quite often the word “disruption” comes up when talking about new technologies or digitalization which is, in a sense, regrettable because “disruption” has somehow a negative connotation: who wants to disrupt his business?

Along the same lines, sometimes we can read or hear that the technology is the driver of the shipping industry transformation. Again, one can wonder, is that so? Shouldn’t it be the other way around? Shouldn’t the technology be the one driven by the industry needs?

And the final question, what is the role of a Classification Society in this transformation?

Starting with the last question, as a Classification Society our role is not to promote one technology or another, our role is to provide a comprehensive regulatory framework to enable the use of these new technologies in a safely manner for those who want to take advantage of them. For that, all the potential additional risks are identified and efficient mitigation measures are defined in form of rules or guidelines in order to ensure that our clients can take safely advantage of the new technologies.

Connected sensor is just an example of where the industry is heading to. Connected sensors will change the traditional survey scheme required by classification societies. Indeed, the use of connected sensors will enable inspections and maintenance based on the data and based on the degradation model of the equipment instead of inspection at fixed intervals, it will prevent the “too soon, too late” risks of the traditional preventive approach and reduce the consequent additional costs due to downtime and unplanned maintenance.

As a part of the digital transformation, systems on board are more and more controlled by computerized control systems and that including essential safety system. By publishing the Unified Requirement E22, the International Association of Classification Societies (IACS) has acknowledged the need for having appropriate requirements to ensure that the computer systems fulfil their intended function. Bureau Veritas has fully implemented these requirements together with new notations such as HWIL (Hardware in the Loop) which provides a comprehensive framework to safely carry out testing of control systems in a simulated environment.

However, as much as Bureau Veritas sees connectivity, data analytics and the increased use of computerized systems as a great opportunity to improve safety, reliability and to reduce costs, it also implies new cyber risks. To prevent these cyber risks, Bureau Veritas is able to identify, analyse, assess risk and certify that the appropriate mitigation measures have been implemented. Ships which comply with these requirements will by assigned the service notation CYBER MANAGED, when the security control is based on manual procedures or CYBER SECURE when the security control is ensured by means of automatic software.

In parallel, IMO has recognized the urgent need to raise awareness on cyber risk threats and vulnerabilities by adopting the IMO Resolution MSC.428(98) and setting a deadline of 2021 to incorporate cyber risks into the safety management systems.

What about the future?

The increase of computerized control systems on board combined with connectivity certainly takes traditional shipping on a path towards autonomous shipping. To support and assist yards and owners to enhance the autonomy of their ship, Bureau Veritas published in December 2017, “Guidelines for autonomous shipping”.

Even more, a good indicator showing that the industry is moving in this direction is the fact that IMO has officially started to look into Maritime Autonomous Surface Ships (MASS) operations. The Maritime Safety Committee

(MSC) has agreed on a framework for a regulatory scoping exercises and has established a correspondence group. There are two main objectives as far as the regulatory scoping exercises is concerned. The first objective is to define clearly the different degrees of autonomy depending on the role of the human element versus the role of the technology. Secondly, the scoping exercises aims to assess which are the current IMO instruments that may need to be amended in order to integrate MASS operations. In addition, the correspondence group was instructed to carry out a test of the framework and methodology by conducting an assessment of SOLAS Regulation III/17-1 (Recovery of persons from the water). Following the test, MSC 100 approved the framework and methodology so now IMO is ready to move to the second step: analyse and determine the most appropriate way to integrate the MASS operation into IMO tools (develop new instruments, amend the existing ones…)

6. CONCLUSIONS

The conclusion of this paper can be summarized in one word: sustainability. First of all, the increased awareness of climate change promotes the research and development of alternative sources of energy in order to reduce our dependency on fossil energy, in other words, in order to ensure the sustainability of our economic and social systems without compromising the future of the next generations. In this context offshore wind industry has every chance to become a key contributor to achieve sustainable development goals.

However, offshore wind evolves in a competitive market. Primary, offshore wind needs to fight back against other sources of energy so cost reduction creates a lot of pressure even more now that the subsidy landscape is changing. This pressure is applied to the complete supply chain, including the ships supporting the industry. Again, sustainability is the key word, it is essential to find ways to keep the industry sustainable while reducing significantly the costs. Embracing new technologies can definitively help to improve safety, reliability and reduce costs. And this is where Classification Societies and International Organisation as IMO have a role to play; the risks associated to these new technologies are to be identified, analysed, assessed and appropriate mitigation measures are to be defined in the form of rules and guidelines in order to ensure that the new technologies are implemented safely on board.

7. AUTHORS BIOGRAPHY

Eva Peño is the Global Market Leader for offshore service vessels (OSV) and tugs at Bureau Veritas. As a member of the commercial department team of the Paris headquartered classification society she is leading the international business development activities, driving the implementation of the key account strategy and steering the development of new services and standards.

Eva joined Bureau Veritas in 2001 after obtaining a degree of Nuclear Technology from the Escuela Técnica Superior de Ingenieros Navales in Madrid and has since held various technical and project management positions in Madrid and Paris.

.

DESIGN OF AN INNOVATIVE AUTONOMOUS VESSEL TO SUPPORT WINDFARM O&M

Akhil Nair, Arun Pillai, David Pugh, Luiz Demenicis & Marta Więcka, Houlder Ltd. U.K.

SUMMARY

Wind Farm Operations and Maintenance has challenges in delivering cargo safely and efficiently and with manning in high sea states. Wind farms farther offshore further increases pressures on vessel design, technicians, safety and costs.

Working within the ambit of 'Innovate UK' funded Wind Farm Autonomous Ship Project, Houlder Ltd have been developing the concept design of a Wind Farm Support Vessel that can operate autonomously during transit and while delivering cargo to offshore wind turbine. The aspired degree of automation is that for the vessel to be remotely operated from ashore. Vessel functions have been identified, working in conjunction with industry and project partners, who are developing vessel control and advanced autonomy, on-board health diagnostics, decision support logistics algorithms, data flow and logistics management, and cost performance analysis, as part of the project. The Concept is expected to reduce operating costs, especially fuel and crew related, while improving safety of operations.

This paper will focus on the re-think, re-design and refining of vessel design that the autonomous vessel concept has accorded. The discussion will follow through the design spiral for features of spaces, stability and machinery systems, besides the innovation and integration of autonomous cargo handling solution for the wind farm O&M.

NOMENCLATURE

ASV	*Autonomous Surface Vessel*
CTS	*Cargo Transfer System*
CTV	*Crew Transfer Vessel*
COLREGS	*International Regulations for Preventing Collisions at Sea*
DoF	*Degree of Freedom*
GPS	*Global Positioning System*
IMO	*International Maritime Organisation*
IS	*International Standard*
IT	*Information Technology*
LCoE	*Levelised Cost of Energy*
O&M	*Operation and Maintenance*
OWE	*Offshore Wind Energy*
OWF	*Offshore Wind Farm*
OWT	*Offshore Wind Turbine*
RAO	*Response Amplitude Operator*
RAI	*Robotics and Artificial Intelligence*
SOLAS	*Safety of Life at Sea*
SOV	*Service Operating Vessel*
Te	*Tonne*
UK	*United Kingdom*
WAS	*Windfarm Autonomous Ship*
WASP	*Wind Farm Autonomous Ship Project*
WTG	*Wind Turbine Generator*

1. INTRODUCTION

Renewable Energy, led by solar and wind, are increasingly contributing to electricity generation worldwide [1]. Correspondingly, innovations in IT, machine learning and sensor technology have led to early adoption of autonomous system in land, air and marine transportation.

Offshore wind farms are heavily dependent on human intervention across construction, installation and O&M project stages. UK Government's Industrial Strategy funding for RAI has facilitated identification of autonomous technologies for potential use with data collection, inspection, cargo delivery and security in the sector. Working with Innovate UK [2], Houlder Ltd have leveraged on their vessel design and motion compensated marine equipment competencies to develop the concept of an autonomous vessel of significant size (about 24m) with an integrated autonomous cargo handling system.

This paper discusses the vessel design aspects of the innovative WAS, supporting planned maintenance of the offshore wind farm.

1.1 BACKGROUND

Barely in existence ten years ago, OWE industry comprised over thirty wind farms around UK shores in 2017. UK now has the largest installed wind capacity in Europe (43% of installed capacity) and is the largest OWE market globally. In 2016, The UK Government forecast OWE to provide 10% of the UK electricity demand by 2020. The number of turbines is predicted to be over 5,500 by 2025 [3]. Within the industry, windfarm development and construction have been the focus of attention in the UK for many years. With many assets commissioned and numbers growing, O&M now provide increasing technical and commercial opportunities.

'Operations', in the OWE context, refers to asset management activities such as remote monitoring, environmental monitoring, electricity sales, marketing, administration and other back office tasks. 'Maintenance', comprising preventative and corrective variants, is the 'upkeep and repair of the physical plant and systems', and accounts for largest portion of effort, cost and risk. The key concepts shaping the O&M services are availability, scheduled and unscheduled maintenance, access and cost reduction. Increasing accessibility is seen to increase turbine availability and uptime. Increased workboat speeds & passenger comfort increase fuel consumption and place a limit of 4-hour one-way transit time to facilitate an 8-hour maintenance activity in the field. However, additional safety regulations and vessel design changes offset advantage and calls for innovative solutions. Further, improving fuel economy will reduce O&M costs, as fuel approximately accounts for 30% of vessel operating budget [4].

Marine logistics impacts O&M cost with its dependency on the weather window and capability to deliver cargo. Presently the O&M market is served through various marine fleet such as Crew Transfer, Service Operation and Jack up vessels. A baseline review of current operational scenarios and future autonomous scenarios carried out [2] has identified five use case scenarios for marine logistics:

- cargo delivery to SOV
- cargo delivery from SOV to OWF
- geophysical survey
- security operation and
- in-field crew and cargo transfer.

Operational statistics show that human error, especially during cargo handling, has been the primary cause of marine accidents [3]. Decoupling humans from the cargo delivery function has the potential to improve safety, accessibility and efficiency.

Autonomy is not wholly unfamiliar to the marine sector, as container feeder vessel Yara Birkland is planned to move from manned to fully autonomously operation between 2020 and 2022 [5]. Rolls Royce led AAWA (Advanced Autonomous Waterborne Applications) project specifically aims at the design of ocean-going autonomous ship. Furthermore, MUNIN (Maritime Unmanned Navigation through Intelligence Network) project funded by the European Commission has developed an autonomous ship concept. In 2018, Finferries and Rolls-Royce successfully trialled autodocking and navigation of ferry Falco with people onboard [6].

The WAS concept will be developed working in conjunction with industry and WASP project partners, who are developing vessel control and advanced autonomy, on-board health diagnostics, decision support logistics algorithms, data flow and logistics management. Cost performance analysis will develop LCoE model around turbine availability linked to sea state variability, distance from shore, fuel costs and vessel maintenance.

The design philosophy and basis for the vessel therefore conforms to that of a cargo vessel remotely controlled from shore with the primary role of small cargo delivery to the wind farm for the preventative and scheduled maintenance related to the O&M activity, with the possibility of carrying out secondary roles such as survey and security.

2. DESIGN PHILOSOPHY

WAS has derived functions from the singular user need to improve safety, accessibility, and efficiency of cargo delivery for O&M operations. No significant change to the existing OWF infrastructure is envisaged in developing the concept. Further, keeping with the aspirations of the industry partner, WAS looks to deliver cargo directly to the OWT, without recourse to any existing lifting devices on the OWT.

WAS intends to load the cargo from an onshore O&M base and deliver it to the offshore OWT. Delivery of cargo to the maintenance island or infield SOV is not considered. WAS service takes a planned route for travel from O&M base, is remotely operated from shore and is fitted with a CTS to automate the process of loading and unloading cargo. The cargo delivered is in 1m^3 containerized units, weighing not more than 1 tonne, targeted at the lowest cargo sets for the planned maintenance operations of the wind farm.

WAS adopts the traditional ship design spiral (Figure 1) to translate the key user requirements to a concept design fit for development of feasibility and detailed design. The design has looped through size, shape, volumes, power, machinery, stability and seakeeping considerations to realise the concept. Within the spiral, the innovation of remote-controlled operation has accentuated considerations on reliability, redundancy and maintainability, thus simplifying deck layout and refining volume arrangement.

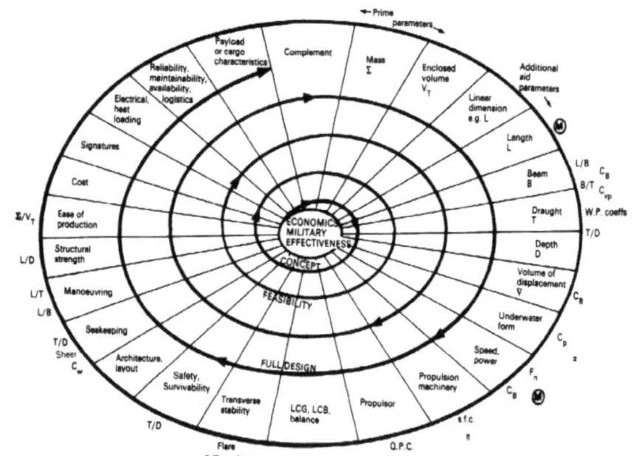

Figure 1: Traditional Ship Design Spiral [7]

Autonomous vessel operation brings concerns on the system break-down with no human on board to provide interface between the mechanical system and controls. Simplicity of design and the redundancy of critical systems therefore becomes important. Where the design solutions presented a choice between simpler adaptation of existing provision and adoption of complex automation, the former has been chosen. Also, simple processes have been considered for autonomous systems adopted. Further, reliability of the systems facilitates uninterrupted operation. However, as a contingency plan, provisioning of human access, ventilation and escape routes becomes necessary. Redundancy, reliability, simplicity and maintenance considerations have been the drivers for WAS vessel design (Figure 2).

In applying existing rules and regulations for the design, the absence of personnel onboard on WAS during operation calls for pragmatic interpretation. SOLAS considerations of access, egress and fire safety are cases in point. Based on the unmanned operation, provision of these considerations may be minimised. However, personnel intervention at sea required for maintenance and obligations of WAS to life at sea, assets and environment pose a paradox. While COLREGS provides technology challenges for WAS in the dynamic autonomy function of navigation, manoeuvring, mooring and anchoring its impact on the static function of design is not considered ground-breaking.

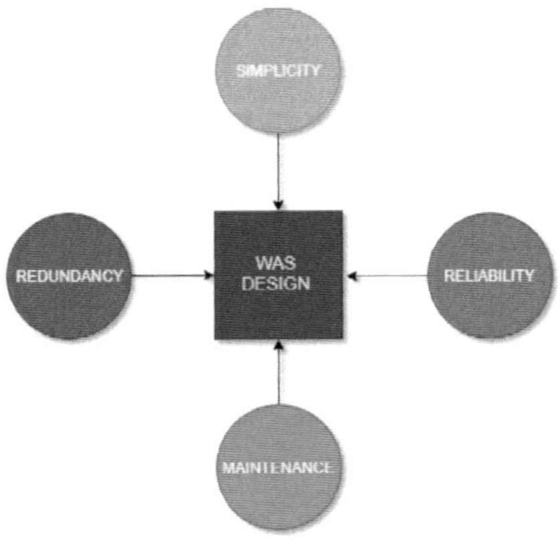

Figure 2: Design Drivers for WAS

The prescriptive, rule-based design has accorded a minimum level of safety for design of conventional ships over the years. Rules and regulations for autonomy, defining high level requirements will take time to evolve, although scoping exercises have been initiated by IMO and select Flag Administrations. Guidelines on technical assurance, a function long provided to the maritime industry by classification societies, will follow the high-level regulatory requirements on design and operation. In the meantime, adoption of innovation through remote-controlled systems and operations will require assessment of risk vis-à-vis events, likelihood and impact on the design and operation with a view to maintain at least similar levels of safety to the existing paradigm. It is possible that the regulation of safety assurance in the industry will adopt a goal-based design approach, akin to the oil & gas sector, for the autonomous ships.

WAS has adopted the above design philosophy in that hazards relating to each aspect of design and unmanned operation has been assessed for its impact on design. Identifying related hazards, in general and specific to WAS, prioritising these and engineering design solutions to avoid, control, mitigate, modify, minimise or accept design associated with the risks have been part of the design process. Risks of collision and loss of control are examples of high risk. Spaces, machinery and system concepts have been re-configured on such basis.

3. CONCEPT DESIGN

3.1 SPACE

The spatial arrangement of the vessel has provided for the twin functions of wind farm and autonomous operation.

The displacement of the vessel has been selected to conform to be just above that of CTVs, keeping within the industry partners' preference of vessel size that will service the WTG for O&M in the near future.

WAS delivers the cargo to the transition piece platform positioned at 15-20meter above mean sea level. The intended delivery from the vessel poses design challenges but will preclude use of and interface with WTG lifting infrastructure.

A single tier of the cargo is laid out on the main deck (Figure 3). It is envisaged that cell guides will sea fasten the cargo. Standardised cargo units facilitate cargo handling automation and weathertight cargo pallets that weigh up to one tonne each having one square metre foot-print have been considered. Euro pallet sizes have also been checked to confirm that these can be accommodated within the designated cargo area on deck, in lieu of the one square metre foot print container currently considered.

WAS is not constrained by manning comfort and shift hours and the vessel speed therefore becomes non-critical for the operation. Consequently, the speed has been reduced for vessel transit, with resulting reduction in power requirement and machinery space.

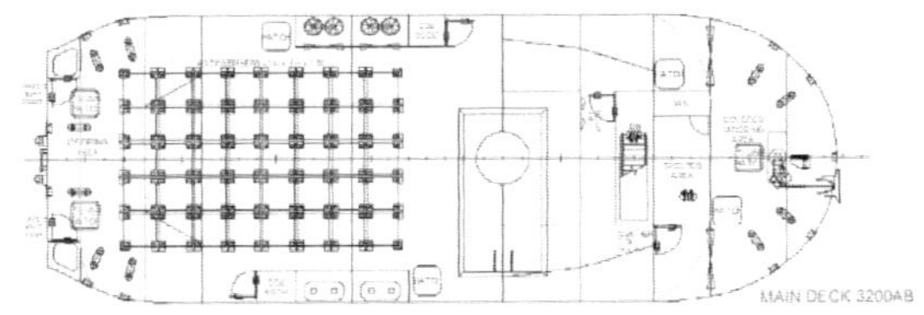

Figure 3: WAS Main Deck Layout

The transit and manoeuvring are achieved through azimuth thrusters – these also help with station keeping alongside the two bow thrusters. As the vessel is beamy (low length to beam ratio), skegs are provisioned to improve course keeping, while also providing additional stability by way of lower vertical centre of gravity and improved roll motion damping.

WAS requires sensors, GPS and an air-conditioned control equipment room for autonomous navigation and cargo handling.

The vessel relies on the redundancy of the systems ensuring safe return to the port. The engine rooms, propulsion system and bow thrusters are provided such that a single compartment flood or single system failure does not interrupt the vessel operation (Figure 4).

Access and escape routes are provided for maintenance. Human access for the maintenance to the ship is given through the boat landing located at the stern. Health monitoring system onboard for propulsion components for early detection of malfunction and preventative maintenance is also envisaged and parameters defined.

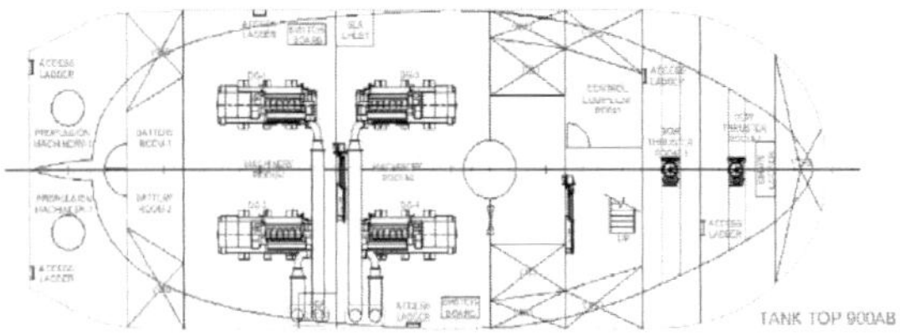

Figure 4: WAS Tank Top Layout

The auto mooring system would necessitate revamping the port infrastructure for enabling auto docking such as Moormaster system. Besides, the vessel is equipped with anchor arrangement for use in case of emergency situations.

Internal ballast operations are necessary for trim and heel corrections due to a shift in the centre of gravity of the operating crane arm and an asymmetrical 'container' cargo loading about the centre line.

3.2 STABILITY

Following IS Code criteria have been analysed for WAS design:

- general criteria regarding lever curve properties [8]
- severe wind and rolling criteria [8]
- ships engaged in lifting operations criteria [9]
- sudden loss of hook load [9].

It is evident that existing, applicable stability regulations are not easily amenable for interpretation to suit unmanned operation. IS Code explanatory notes [10] ascribe the origin of the present stability criteria to analysis of stability casualty records and statistical data gathered for vessels which have been lost, heeled and safely operated. A dearth of autonomous vessel operational data prohibits any exemption of criteria on the basis of unmanned operation.

As per the IMO criteria for vessels engaged in lifting operations, the cargo handling function has been investigated for every lifting condition during the operation and transit of the vessel. Maximum static heel, trim angle, draught and propeller immersion have been assessed. The intact stability analysis has shown that the cargo must be unloaded symmetrically for good stability during the lifting and cargo delivery operation. Transverse crane operability range has been limited on the vessel to maximise the heel angle at which autonomous crane can operate.

WAS complies with the IS Code requirement for ships engaged in lifting operations to withstand sudden loss of hook load and use of counter ballasting. Fresh water ballast counters the heel and trim associated with field transit and cargo handling.

The unmanned operation of the vessel inhibits provisioning the ventilation openings on the main deck with conventional weathertight covers. To accord simplicity of design, in lieu of automated weathertight closures, these openings have been raised and considered as unprotected in the stability calculations. The autonomous vessel function and operation also accord flexibility in adjusting the position and height of the openings.

Provision of skegs have further facilitated conformance to all stability criteria and design requirements. Hull shape has been modified by introducing three skegs below the base line to provide buoyancy and modify the longitudinal centre of gravity to suit WAS operation (Figure 5).

The design for damage stability is based on the Guidelines for the Design and Construction of Offshore Supply Vessels [11,12] as there is no evidence of equivalent criteria in local authority regulations.

Subdivision has been modified to reduce lost buoyancy and limit the damage penetration to two compartments. It is considered feasible to increase the prescribed restriction on the heel angle due to unsymmetrical flooding on the basis of unmanned operation. However, it requires that the resultant heel angle is investigated during every crane operational and transit condition.

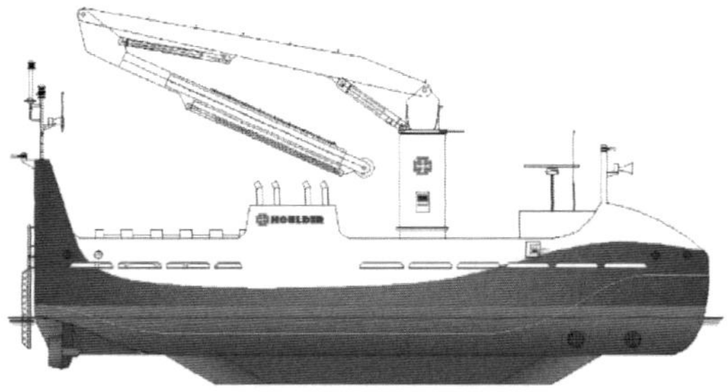

Figure 5: WAS Profile

3.3 CARGO HANDLING SYSTEM

An innovative crane handling system delivers the cargo from WAS to the turbine transition piece platform. The delivery height above main deck translates small motion amplitudes to large displacements at the crane tip. Operational requirement has precluded crane contact with the turbine and this constraint has added to the complexity of design. A multiple DoF motion compensated crane using hydraulic actuators (mix of cylinders and/or Motors) is therefore required.

The primary objective has been to evaluate the feasibility of reaching the delivery point while keeping the position of the crane tip stabilised immediately above it. The telescopic lowering device is to be developed.

The design provides means of compensating for three DoF using three actuated joints: slewing, folding, and extending functions (Figure 6), while a fourth cylinder locks the luffing boom at a pre-set position. These three DoF allow the crane to keep its tip at the same position in space during the vessel motion, while the luffing cylinder can also act as an emergency pull-away device. Additionally, four cylinders can be used for general handling of goods anywhere on the vessel deck.

Houlder uses its experience of developing 18m knuckle boom crane to extend the concept of using standard crane as a motion compensated, autonomous handling device for delivering one tonne containerised cargo on the OWT transition piece platform.

Parameters such as vessel motions and size, equipment weight, cylinder forces and overall power consumption are benchmarked against similar conventional crane designs to gain confidence on the concept feasibility. The cylinders displacements, loads and strokes are studied for the operating sea states. The analysis considered range of wave periods, wave directions and vessel motion RAO.

Cylinder displacements for full motion compensation are established assuming the boom tip fixed in space while the vessel is subjected to pre-calculated motions. These displacements input for each of the actuators derive the cylinder loads and the overall power consumption. An accidental load case, from locking out of the cylinders has further been studied. It is noted that the crane imparts a coupled motion to the vessel that results in high roll motions in the beam sea, threatening vessel stability. The coupling between the crane and vessel has therefore required design iterations on vessel proportion and operability.

The strength of the crane is checked for the accidental case, while the maximum displacements are checked for the operating case. Operating and accidental design cases based on wave headings are also intended to be derived.

Considerations of crane stiffness and natural frequency, effects of friction and viscosity and contingency for a DP watch circle needs investigation. Further, control system for operation and cargo landing and release needs looking at - however these are not considered critical to the concept design.

Operating envelopes greater than 2m H_s have been investigated and are possible.

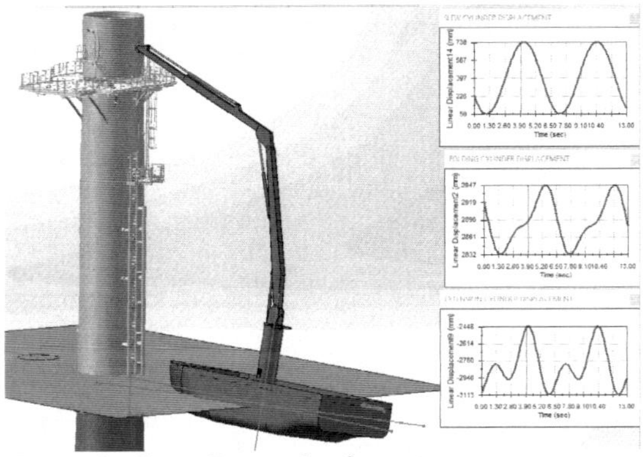

Figure 6: Crane Design Configuration

3.4 MACHINERY SYSTEM ARCHITECTURE

The key drivers for the propulsion system are the vessel reliability and ability to deliver cargo successfully to the wind turbine up to the maximum operating conditions envisaged.

As an autonomous vessel, no crew are present to provide planned or corrective maintenance whilst the vessel is deployed with the only maintenance tasks being carried out whilst the vessel is in harbour. Therefore, a diesel electric propulsion system has been selected due to the high level of redundancy that can be applied, including multiple and distributed generator sets, electrical distribution equipment and propulsors. The conceptual machinery system architecture is depicted in the figure 7.

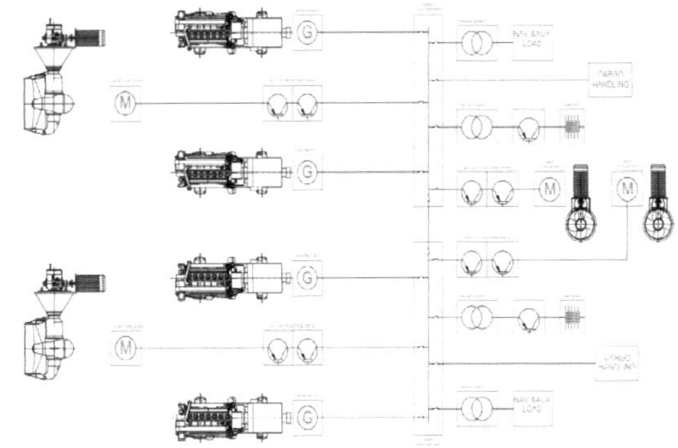

Figure 7: Propulsion System Architecture

To successfully deliver cargo the vessel must maintain station within a defined set of parameters to suit the cargo handling equipment, and hence Dynamic Positioning (DP) is required. The category of DP system defines the level or redundancy and the vessels ability to keep station in the event of a failure. When considering the category of DP system, the consequence of a DP failure resulting in a collision with the wind turbine is the key factor. In carrying out the design, the impact energy of the vessel with the wind turbine was considered and kept below a level that represented a risk to the wind turbine, therefore negating the need for a highly sophisticated DP system.

As with the propulsion system, the key design drivers for the auxiliary systems is vessel reliability. This is provided by

- selecting design solutions based on reliability
- providing duplication of essential system components to allow continued operation in the case of a single point failure
- simplifying the design to remove all nonessential features.

For example, box coolers were selected to provide cooling to the diesel generators, eliminating the need for a sea water cooling system and the failure risk and maintenance associated with sea water and marine fouling.

4. CONCLUSIONS

The growth of renewables in the energy mix coupled with the large number of offshore wind farms around UK shores present opportunities in wind farm O&M to the marine industry. Alongside this, the transportation sector has been early adopters of innovations in automation, driven by IT and sensor technology. These market and technology indicators have driven the Innovate UK funded concept design of a Windfarm Autonomous Ship at Houlder.

With the overall aim of improving the current O&M logistics model, the WAS concept derives functions to enhance turbine accessibility, vessel operability and improve safety and efficiency. The WAS services planned maintenance and is suited for transporting and delivering cargo to the offshore wind farm turbine, remotely operated from shore.

The WAS design philosophy is driven by the need for simplicity, redundancy, reliability and maintenance. Rules and regulations for marine autonomy are currently evolving and a risk-based design approach is adopted for the design of spaces, stability, propulsion, fire, ballast and cargo handling systems.

The WAS general arrangement rethinks onboard spaces by eliminating accommodation, providing for emergency maintenance at sea, simplifying access and modifying subdivision for enhanced redundancy.

Intact and damage stability assessments have ensured that the WAS maintains its buoyancy, stability, weathertight and watertight integrity to a level of safety equivalent to that of a conventional cargo handling vessel.

The WAS integrates an innovative cargo handling crane concept that reaches the turbine platform to deliver small units of cargo for O&M. The full motion compensating knuckle boom crane concept design parameters are benchmarked against conventional crane designs for feasibility validation.

The propulsion and auxiliary system designs have been driven by the need for reliability, redundancy and simplicity demanded by the unmanned operation and lack of potential corrective maintenance onboard. Diesel electric propulsion, DP positioning and elimination of sea water cooling are cases in point.

Operating envelopes greater than 2m H_s for WAS operation have been investigated and are possible.

5. ACKNOWLEDGEMENTS

We acknowledge Innovate UK for their funding of the WASP and project partners L3 ASV, University of Portsmouth, SeaPlanner Ltd and ORE Catapult for their contribution and sharing of expertise on developing and interfacing on vessel control and advanced autonomy, onboard health diagnostics, decision support logistics algorithms, data flow and logistics management, and cost performance analysis respectively.

6. REFERENCES

1. BP Energy Outlook 2018 Edition.
2. IUK *Windfarm autonomous ship project.* https://www.gov.uk/government/organisations/innovate-uk, 2018.
3. *Offshore Wind - Operational Report.* The Crown Estate, 2017.
4. GL GH, 2013. A Guide to UK Offshore Wind Operations and Maintenance. London: Scottish Enterprise and The Crown Estate.
5. Ship & Boat International 'Yara Birkeland due in 2020' eNews: September/October 2018.
6. Marine Propulsion and Auxiliary Machinery 'Rolls-Royce and Finferries successfully trial world's first autonomous ferry' December/January 2018/2019
7. Basic Ship Theory (K.J. Rawson and E.C. Tupper)
8. Resolution MCS.267(85) Adoption of the International Code of Intact Stability, 2008
9. Resolution MCS.415(97) Amendments to Part B of the International Code on Intact Stability, 2008 (2008 IS Code).
10. MSC.1 Circ.1281 Explanatory Noted to the International Code of Intact Stability, 2008.
11. Resolution MCS.235(82) – Adoption of Guidelines for the Design and Construction of Offshore Supply Vessels, 2006.
12. Resolution MSC.335(90) - Adoption of Guidelines for the Design and Construction of Offshore Supply Vessels, 2006.

9. AUTHORS BIOGRAPHY

Akhil Nair is a Naval Architect at Houlder. He is responsible for conceptual design of vessels. His previous experience includes design of commercial and naval vessels.

Arun Pillai is a Principal Naval Architect at Houlder and a member of RINA. He manages projects, including WASP, on concept and detailed design of floating structures. His previous experience includes design, analysis, R&D and assurance of structures in marine and offshore sectors.

David Pugh is a Principal Marine Engineer at Houlder Ltd and a member of IMechE. He is responsible for main machinery and auxiliary system during vessel concept and detailed designs. His previous experience includes commercial and naval vessels including a windfarm installation vessel, passenger ferries, offshore patrol vessels and cargo ships.

Luiz Demenicis is a Principal Mechanical Engineer at Houlder. He is responsible for concept and detailed designs of equipment. His previous experience includes concept-to-completion of pile gripper arms, pin pile gripper, monopile lifting tools, fibre rope winch and pile guides for subsea pile template.

Marta Więcka is a Naval Architect at Houlder and member of RINA. She is responsible for stability calculations. Her previous experience includes stability calculations for various ship types mainly using Napa software.

CTV operational experience from 7 to 27m with stern drives and pod drives

G. Torneman, AB Volvo Penta, Sweden

SUMMARY

The distance to the offshore windfarm sites as well as the size of installation and support vessels are constantly increasing. Today you can find vessels up to 100m, all vessels being perfectly suited for its tailored task.
This paper will present Crew Transfer boats, ranging from 7m to 27m, purpose built for its operation and designed for efficient crew transfer, from shore or from hotel ships to windfarm structures, platforms or special operation vessels. The paper will focus on vessels using active drive systems as sterndrives or pod drives, showing how they work safely and efficiently in crew transit and crew transfer operations.
With contra rotating stern drives and contra rotating pod drives the active maneuvering in high seas and the high bollard push against any boarding structure will ensure that the downtime due to bad weather is minimized and the safety for the crew and technicians are kept on a high level. The CR technology will ensure that vessels are keeping their grip in the water and at the same time showing up to 35% lower fuel consumption
Monohull and catamaran vessels will be presented together with selected propulsion systems and experience from real operation in different duties will be shown.
The paper will give a good insight in the performance and the flexibility of the stern drive and pod drive propulsion system in different applications, with over 10 years of experience in offshore wind environment. It will also show the future of an electro mobility concept for parallel hybrids in this kind of environment.

NOMENCLATURE

CTV	Crew Transfer Vessel
WFSV	Wind Farm Support Vessel
SOV	Special Operations Vessel
STB	Safe Transfer Boat
FRB	Fast Rescue Boat
IPS	Inboard Performance System
EVC	Electronic Vessel Control
FPP	Fixed Pitch Propeller
CPP	Controllable Pitch Propeller
WJ	Water Jet propulsion system
CR	Contra Rotating propellers
DUOPROP	Contra Rotating prop on drive leg

1. INTRODUCTION

The origins of Volvo Penta can be traced back as far as 1868, when the foundry and mechanical workshop Sköfde Gjuteri & Mekaniska Verkstad was founded. The company manufactured cast-iron goods and hardware such as boilers, pots and pans, ploughs and threshing machines. A somewhat humble beginning for a company that has become a world leader in diesel engines for marine and industrial use. The 50s was a decade of rapid development in products and business for Penta, and the company developed into a truly international operation with an extensive network of dealers all around the globe. 1959 saw the introduction of one of the first Penta products to take the market by storm – the Aquamatic sterndrive. The new millennium sees a continued focus on development of user-friendly and performance-enhancing technology. In 2003 the Electronic Vessel Control is launched and in 2005, the IPS is unveiled and subsequently complemented the very next year with the introduction of the Joystick.

The company is renowned throughout the world for innovations such as Volvo Penta IPS propulsion system and the Aquamatic sterndrives. The Aquamatic is considered to be the most important 20th century innovation in the boating world. This paper is mainly covering applications using Volvo Penta engines. It's based on discussions with designers, yards and operator in the offshore wind segment and on Volvo Pentas own experience in the field. Any statements should be seen as opinions solely from the author, if not been based on real case studies.

2. VOLVO PENTA PROPULSION SYSTEMS

2.1 WORK BOAT SUPPLIER

Volvo Penta has been a leading supplier of engines, marine gen sets and propulsion system to the offshore work boat market ever since the need for efficient crew transfer and support vessels started to grow. With a broad range of products in the power band from 80kW up to 662 kW, covering power and propulsions applications with single, twin and multiple screw combinations, Volvo Penta has taken a lead as a supplier of the most fuel efficient and safest propulsions system in the market. With focus not only on supplying fuel-efficient engines, Volvo Penta has taken an innovative lead in supplying complete propulsion system, from prop to helm, to ensure that all components in the drive train are matched and optimized to each other, in order to enhance performance and reduce the cost of ownership

2.2 DUOPROP CONTRA ROTATING PROP

As this paper will focus on applications with pod drives and stern drives, a small introduction to the proven success of the DUOPOP technology is in place. In 1983

***** Formatting Issue - Best Available Paper/Graphic *****

the DUOPROP was introduced, marking a significant step forward for the Aquamatic sterndrive concept. Two contra rotating propellers on the same drive are giving a greater efficiency and better handling characteristics compared with single propellers. The DUOPROP concept resulted in a much-improved grip in the water, giving better acceleration and unrivalled handling. Both at low and high speeds, a boat equipped with DUOPROP is easy and safe to manoeuvre, the drive leg being an active rudder instead of a passive one. The basic reasons for the high efficiency of the DUOPROP contra rotating technology are showed in fig 2.2.1

Figure 2.2 1 Duoprop vs Single prop

2.3 RAPID GROWTH IN OFFSHORE WIND

Volvo Penta engines were introduced to wind farm workboats in the 7-18m range through early cooperation with Windcat Workboats, Northern Offshore Services and Esvagt, all pioneers in new fresh approach to efficient crew transfer at the offshore windfarms. Volvo Penta has during the last 15 years gained vast experience of different propulsion system such as WJ, FPP and CPP in different crew and work boat applications and in demanding operations offshore. Today Volvo Penta engines and drive trains are used in approx. 15% of the offshore wind farm workboats today in the North Sea, Baltic Sea, the Channel and the Irish Sea.

2.4 VOLVO PENTA IPS IN CTVs and WFSVs

Northern Offshore Services, SIMA Charter, CWIND, Njord Offshore, Opielok Offshore Carriers and Esvagt are some of the offshore energy service providers using Volvo Penta IPS pod drives for their crew transfer and support vessels. Northern Offshore Services (NOS) started already during 2009 to do extensive field testing with commercial rated twin IPS in South Boats 16-18 m catamarans. Today the vessels are running successfully around 3000 hour per year and are always the first ones out on the fields in rough weather due to high manoeuvrability in heavy seas and high bollard push, ensuring safe transfer of technicians to and from the turbine ladder in significant wave heights up to 1.5m. The IPS system is now used both in mono hulls and catamarans and is creating additional course stability of the aft ship when cruising in following seas.

3. SELECTION OF PROPULSION SYSTEM

3.1 FOR 7-17M CREW BOATS

This size of vessels can accommodate 5-12 persons plus crew. The most common propulsion systems used today for smaller crew boats and transfer boats are sterndrives, waterjets and pod drives powered by diesel inboard engines. Majority of the smaller crew boats are mono hulls and most of them can be lifted in cranes or davits, alternatively in special designs of launch and recovery system designed for newer SOVs. Outboard engines are not considered in this presentation due to their high fuel consumption compared to inboard diesel engines, their lack of catalysts and their short time between overhaul. The preferred propulsion systems for smaller crew boats are sketched in Figure 3.1.1 (ref TUCO Marine)

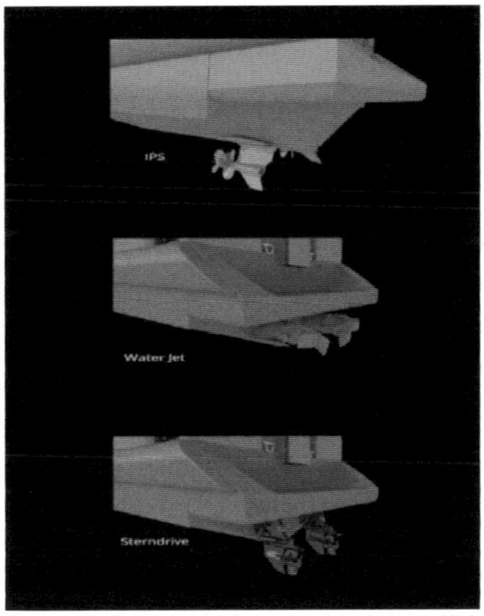

Figure 3.1.1 Propulsion systems for small CTVs

3.2 FOR 18-27M CREW BOATS

In these range (with a load line length less than 24 m) over 90% of the vessels are of catamaran type, to create an efficient work space and safe loading platform for the crew and the operators. Today's owners of larger crew transfer and offshore wind work boats are looking into fuel cost, carbon impact and lifecycle costs when putting up the design requirements for their new vessels. Catamarans with slender hulls can be very fuel efficient at semi planning speeds in moderate seas, with good comfort for the passengers on-board. Therefore the selection of an efficient high performing propulsion system matched with the hull design is one of the most important tasks when designing a high speed CTV, to be able to meet the performance criteria and transfer requirements during tougher offshore weather conditions. The most common propulsion systems for these size of vessels are sketched in Figure 3.2.1 (ref Sea speed / Carbon Thrust)

Figure 3.2.1 Propulsion systems

In vessel sizes above 23 m, multiple engine and drive installations has been becoming increasingly popular. A catamaran with 4 drivelines will have better redundancy, lower fuel cost, easier maintenance and service compared to a twin installation for same total installed power. An additional important feature is the flexibility in using the number of engines required for each duty. During transit with full load, all 4 engines is normally used to give maximum performance in cruising. During low speed and at idling, only 2 engines needs to be engaged and running, by that saving running hours and reduce total fuel burn.

3.3 REDUCED COST & CARBON FOOTPRINT

It is not only the designer or the yard that decide upon what type of propulsion system the vessel should have, the users of the vessel are also heavily involved, in order to optimize the fuel efficiency and handling of the vessel over the whole duty cycle. Still it's arguing that since the fuel cost is paid by the energy companies, the importance of a fuel-efficient propulsion system is not the first priority for the service providers. What can be more wrong! In the 2017 report the Carbon Thrust offshore wind accelerator (OWA) programme has focused on different vessel parameters to plot the performance of the vessel, but unfortunately not yet the propulsive efficiency, see below quote from the OWA summary ref the project Crew Transfer Vessel (CTV) Performance Plot (P-Plot) Development : "It is important to note that performance is used here to describe primarily vessel motion characteristics, speed and ability to transfer personnel, in different sea conditions and at different headings. It does not cover issues such as vessel construction, system engineering, fuel economy or manoeuvrability".

With increasing fuel prices and increased environmental care, factors like fuel economy and carbon footprint needs to be taken into ago, when measuring the performance of the vessel for wind farm support. The efficiency of operation including fuel consumption per nautical mile, payload per horse power, bollard push per horse power and manoeuvrability in high and low speed needs to be further evaluated, to assist in the development of more cost-effective vessels and increase their capabilities. The selection of the right propulsion system will have a great impact on the efficiency of the vessel, up to 35% higher prop efficiency can be achieved with the combination of pod drives and CR props. In addition, selection of the right fuel can further reduce the CO_2 emissions. Ultimately many of today's CTVs will be retrofitted with a hybrid driveline with electric motors and energy storage system as complement to the combustion engine. Such a parallel hybrid concept will be presented in the end of this paper.

3.4 OWNER & OPERATOR VIEW

From an operator perspective the propulsion system is one of the most important parts for the performance of the vessel and for the cost of operation. Njord Offshore carried out a thorough investigation before they selected the IPS900Quad for their new 26m CTVs in June 2014. The operator had previous experience from crew vessels with FPP.

The main reasons for the operator to select IPS were the manoeuvrability, position holding, bollard push, service support and speed/fuel economy together with room for more cabins and more storage which now has been proven in real life.

3.4 DESIGNER & YARD VIEW

From a yard perspective the biggest advantages of using DUOPROP Aquamatic stern drives and IPS pod drives are the accurate speed predictions (to avoid speed penalties) and that they have one supplier of everything from prop to helm covering, standard propeller set, IPS steerable pod units including gear, drive shafts, exhaust system, diesel engines, cabling, control system, alarm panels, joystick and GPS antenna. It's clear that the yard saves a lot of production time and logistics.

As it is quite tricky to get installation time figures from the yard, we can only go on the quotes from yards and designers, saying that installation time for IPS is half of the time installing other propulsion systems. And you can understand this when you see the list of components needed for an IPS installation compared with other propulsion system. The weight savings using Volvo Penta IPS is between 25-40% and its 50% less components to assemble and install.

4. CREW BOAT REVIEW

4.1 SMALLER CREW BOATS 7-15M

The growth of the large support vessels for offshore wind, such as Installation vessels, Accommodation vessels and SOVs has prompted the need for flexible smaller crew boats, which can transfer crew and technicians from the ship to the turbines as well as from the ship to shore. Different type of daughter crafts launched at stern of the larger vessel was first considered, but with the development of the larger SOVs, Esvagt launched a new concept called STB, Safe

Transfer Boats, see Figure 4.1.1 and Figure 4.1.2. The idea came from the successful story of using converted FRBs for transfer of a smaller number of crew and hand cargo from existing supply vessels and offshore PSVs. With the launch of the first SOV, Esvagt introduced STB7 with sterndrives and STB12 with IPS Pod drives. It was a success and the STB set a new standard for smaller offshore wind CTVs. This concept designed to be used for both MOB and for safe transfer operations has made it possible to use the boats parallel to the W2W gangway system. The flexibility and productivity of this set up is now proven and, in many operations, the total amount of transfers from and to the SOVs has been equally divided between the STBs and the W2W gangways, about 50% of all transfers of personal and cargos from the SOVs at an offshore windfarm have been performed with the STBs. As shown figure 4.1.3, when the vessel is locked into it position with the gangway, the fastest way to get technicians to the other turbines is to use the STB, launched with a crane, with the technicians already sitting in the boat. The STB 7 is powered by two Volvo Penta D3 DUOPROP stern drives and can take up to 5 technicians. The larger STB 12 is powered by two Volvo Penta IPS450 pod drives and can take up to 8 technicians. This size of boats is also used the transfer crew from the SOVs to the harbour up to a distance of 200 Nm. Esvagt has so far performed over 240.000 safe transfers on gangways and boat solutions

Figure 4.1.1 Esvagt STB7 in action (picture from Esvagt)

Figure 4.1.2 Esvagt STB12 in launch position (picture from Esvagt)

Figure 4.1.3 Esvagt SOV with the motion compensated gangway ready, with the STBs ready on deck(picture from Esvagt)

TUCO Marine developed a similar CTVs in 11m range, light weight design in carbon composite to be used on the larger offshore wind support vessels. They have supplied their 11m ProZero Wind Farm Service Vessels to the two Ulstein X-stern vessels ordered by the German based Bernhard Schulte affiliate WINDEA Offshore GmbH. See figure 4.1.3 and figure 4.1.4. The vessels are in charter for Siemens. The first smaller CTVs was designed with Volvo Penta D6 engines and waterjets at operator's request. But today TUCO will recommend pod drives or stern drives for this type of applications. The main reason why not recommending WJ for this type of vessel is the risk of thrust breakdown during bollard push and the risk of broaching in following seas. TUCO has also been successful in delivering smaller CTVs to the Damen shipyard build Wavemaster vessels.

Figure 4.1.3 Windea SOV with Tuco ProZero 11 CTV in davit (picture from Windea)

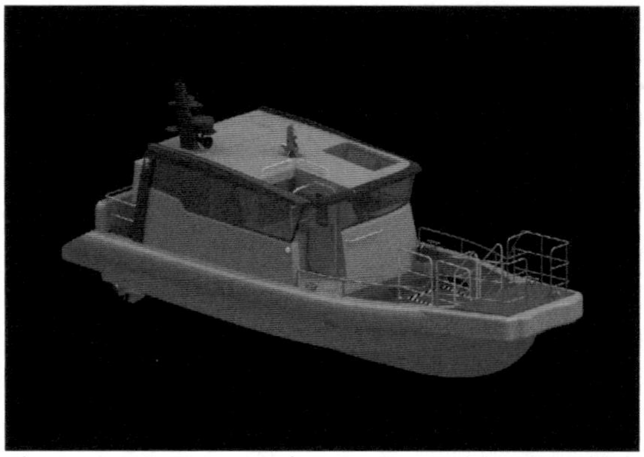

Figure 4.1.4 Tuco Marine ProZero with stern drives (picture from TUCO Marine)

As the SOVs have different design requirements put forward from the operators and the owners, of course also the CTVs will be different. To the Østensjø Rederi, Maritime Partner, being an experienced supplier of FRC, MOB and Daughter Crafts to the offshore industry since 1994, designed their first CTVs, purpose built for offshore wind SOVs. The vessel was similar in shape as the TUCO boats but built in aluminium and designed with Volvo Penta IPS450. Østensjø Rederi first new specialized ship of Rolls-Royce UT 540 WP Design are used for maintenance and service of wind turbines on the "Race Bank Offshore Wind Farm" which is located about 17 nautical miles off the coast of Norfolk, England. The second ship from Østensjø Rederi is working as a service vessel in the Horn Sea wind farm, located about 75 nautical miles off the coast of Yorkshire, England. Maritime Partner will deliver one ALUSAFE 1150 WF and one FRC to each ship. This SOV is shown in figure 4.1.5 while the smaller CTV is shown in figure 4.1.6

The main reason for selecting pod drives for this type of vessel can be summoned in follow quote from Maritime Partner: "While developing the design it has been a priority to ensure seaworthiness, manoeuvrability and operational reliability, while maintaining a high level of comfort for the technicians. ALUSAFE 1150 WF is equipped with a Volvo Penta IPS..propulsion system. Strong manoeuvrability, improved acceleration, user-friendliness and easy service access are amongst the favoured properties of this system".

Figure 4.1.5 Østensjø Rederi SOV with CTV in davit (picture from Østensjø Rederi)

Figure 4.1.6 Alusafe 1150 CTV with IPS450 (picture from Maritime Partner)

Following the operational success of the SOVs and the smaller CTVs and STBs other offshore operators has followed. LDA Louis Dreyfus Armateurs has taken delivery of a new 83m SOV equipped with a new wind farm crew transfer vessel from iXblue in France. This new CTV, named "Speed of Change" will operate on four wind farms off the coast of Germany (Borkum Riffgrund 1&2, Gode Wind 1&2) for transfer of technicians from the SOV to the windfarm sites for the maintenance of the turbines. Made of advanced composites and with LOA of 11.75 meters and a draft of only 0.9 meters, the iXblue CTV can reach a speed in excess of 28 knots and carry 8 technicians and 2 crew members. The design is approved by classification society according to French offshore requirement. Being a light weight carbon composite design with low draft, the Volvo Penta IPS will further enhance the vessels performance by adding good grip in the water in all sea conditions. The active maneuvering with the propeller thrusts and rudder always pointing in the same directions, makes the approach to the turbine landing safe and easy, using joystick docking. Figure 4.1.7 show the Speed of Change in cruising speed of 23 knots

Figure 4.1.7 iXblue Speed of change with IPS450 (picture from iXblue)

The Louis Dreyfus Armateurs SOV can accommodate more than 60 technicians plus crew and is designed by SALT ship design and built by Cemre shipyard in Turkey. See figure 4.1.8

***** Formatting Issue - Best Available Paper/Graphic *****

Figure 4.1.8 LDA SOV with gangway deployed (picture from SALT design)

4.2 LOAD CARRYING CAPABILITY

A smaller CTV should be capable to carry various number of passenger and smaller cargo (tool, parts etc)
The smaller a vessel is the more load sensitive you could say it is. When using pod drives or stern drives a very limited drop in speed will be seen, when carrying extra payload or heading out in rough sea. The increased hull resistance will be compensated by increased thrust, as the contra rotating props on sterndrives and pod drives will just ask for more torque without any slip or cavitation. The waterjets on the other hand will not be able to compensate for the higher resistance and the boat speed will drop, while the engine is maintaining the same rpm. More of this will be discussed in next section.

4.3 LANDING CAPABILITY

As one of the main topics on the OWA programme mentioned has been the access system and various designs has been tested and developed, for the larger crew boats. Some quite complicated and heavy, like smaller motion damping gangways and some simple and straight forward, like the Windcat Windgrip. This rugged and efficient self-tensioning system increases the ability of the vessel to transfer in greater wave heights and reduces fuel consumption whilst pushing on to offshore structures. See figure 4.3.1

Figure 4.3.1 Windcat windgrip in action (picture from Windcat Workboats)

For the smaller STBS life are sometimes even simpler, firstly they cannot carry any additional equipment in fore end of vessel due to added weight in wrong place and secondly, in rough sea, it's safer and faster to rely on a trained helmsman and an experience technician when you climb from the boat to the landing and vice versa. But with the growth of the offshore wind, it is harder and harder to keep on to the trained guys, so you need to employ new untrained land "used" technicians, that you don't want to put into a high risk situation. Then you can say, use the motion damping gangway! But that is not so easy all the time and not so efficient seen from the total operational perspective. So for the smaller vessels two simple systems has been used, one has been to have a sliding fender with moderate bollard push, just to keep the vessel in position but not stopping the vessels bow to move up and down with the waves. This system is for the experience guys and has been used extensively up to now without any real accident, all because of experience crew and passengers. This could be compared to what pilot are doing every day, stepping from a moving platform to a ladder on a ship side, that is something you have to train to do well. To make it easier to step over from boat to offshore structure, push on fenders with high frictions are now tested used more and more for the smaller vessels. The friction between the boats fender and landing structure will keep the landing platform on the vessel in relatively stable conditions and make the stepping from boat to a fixed ladder much simpler compared to stepping out from a heaving platform.

In all landing situation towards a fixed structure or when coming back to the mother vessel, the position control of the CTV is of outmost importance. Pod drives, sterndrives and waterjet drives are all very good system, for active manouvrering and positioning of the vessels towards the landing. Figure 4.3.2 shows how it could look like when you try to position you boats in a situation with high current, you need to be skilled skipper and a joystick docking system could sometime be of good help. Some skippers like to have a bow thrusters, which could make sense for narrow monohulls, where the power from the drives in the stern are not sufficient to rotate the vessels. For a given power, the forward facing CR props from Volvo Penta IPS is giving you the best thrust and acceleration for safe manoeuvres and by vectoring the pod units individually, you can easily side step and rotate the boat with a joystick control system. So no need for bow thruster in this case. The IPS is also giving you a high bollard push, to keep the bow in position when pushing on with a sliding fender or a Push on friction fender.

Figure 4.3.2 Approaching turbine landing (picture from TUCO marine)

4.4 MEDIUM SIZE CTVs 16-22M

The most common CTVs in service for the majority of the already build windfarms in Europe are CTVs with catamaran hulls built in Aluminium and in Composite. Most common propulsion systems has been twin FPP, WJ, CPP and IPS. They are operating in a speed range from 20-30 knots and estimated transit from shore to wind park is not more than 1 hour. Longer journeys in bad weather will put a lot of strain on the passengers and create risk for sea sickness, a condition which is no good idea to be in when arriving to the installation vessel or turbine to do the job. The first vessels equipped with Volvo Penta IPS was three vessels built by South Boats for Northern Offshore Service. They were the boats that went out to the windfarms in so rough weather conditions that made the other service providers to stay in safe harbour. The main reason for the vessels to be able to manoeuvre safely in rough see is the pod drive. Similar catamarans built with WJ had obvious problems to stay in position out at the turbines, the risk of thrust breakdowns was high, due to aeration of the waterjets in big waves. IPS pod drives are less prone to aeration and cavitation problems compared with WJ, due to the fact that props are situated deeper in the water and when the wave surge, the props are still submerged, while the WJ could get air from any side or from transom with reduced thrust as a result. CTRUK / CWIND and SIMA Charter introduced IPS on some of their boat models, to be able to coop with the demand of operation in rough weather conditions. Common at that time (and may still be so) was that if you cannot perform one day's charter your penalty was three days charter rate. And if you cannot perform during a week you lose the contract. So be able to perform in bad weather would always be a bonus. See figure 4.4.1 ref a CTRUK built catamaran with twin IPS900 pushing on the pylon with full bollard pull, keeping the landing platform hinged to the structure through friction. No room and no risk for thrust breakdown in this situation!

Figure 4.4.1 High bollard push needed in heavy seas

The competition in this range of vessels is tough, due to the oversupply of vessels able to work on near shore windfarm. This has created lower day rates, less demand for new vessels in this size and pushing some yards out of business in Europe due to lack of new builds. Most of the vessels in this size are now chartered for the O&M phase of the wind farm and to compensate for lower rates, the service providers offer additional services with their owned skill technicians and craftsmen, for inspection, painting and re fuelling. The company CWIND have had great success with this kind of set up.
Some of the boats has been laid up, but now we have seen some movements with operators starting to retrofit older vessels with new engines and prop systems (instead of overhauling the old engines). The latest now is Turner ICENI in UK, growing their fleet and replacing WJ and engines with new Volvo Penta IPS900, which will give them a CTV with increased performance and reduced emissions. Compared to the WJ, the IPS consumes around 20% less fuel in this vessels speed range with the subsequent reduction of CO_2.
The Volvo Penta IPS pod drives for commercial applications have now been in service for over 8 years with an average number of running hours per year around 2500 hours. Larger vessels are now starting to use the IPS in multiple installations in 24 hour operations, pushing the running hour per year up to 4500 hours. See below some examples of vessels with Volvo Penta IPS pod drives in figure 4.4.2 and figure 4.4.3

Figure 4.4.2 CWIND MPC22 with IPS1050 Twin (picture from Ctruk)

4.4 LARGER SIZE CTVs 23-27M

Development of large offshore wind farms further out from shore has prompted development of a larger range of CTVs, between 23-27m in overall length but still below the 24 m load line length. .Designers like INCAT Crowther, BMT Nigel Gee, Alicat South boat design and now Chartwell design has put forward efficient hulls with good seakeeping for longer voyages, up to 4 hours or maximum 100 Nm from shore for daily services. For such long trips redundant propulsion systems are of great benefit. We have seen propulsion system with 2 engines in each hull coupled to waterjets, FPP and CPP. The CPP solution is normally based on two engines in to a common gearbox with one output shaft. The engines are the mechanically connected to each other. Grovfjord Mekaniska in Norway has delivered 2 vessels of the D-class with 4 Volvo Penta D16M engines and Servogear CPP system to Northern Offshore Services. Those vessels are one of their best performing vessels in the 27m D-range today, see figure 4.4.1

Figure 4.4.1 NOS 27M D-class with 4 x D16MH 750HP (picture from Northern Offshore Services)

Volvo Penta introduced the IPS900 Quad propulsion system based on 2 IPS pod drives in each hull coupled to two high speed diesel engines staggered in each hull The very compact installation creates a very spacious and easy to work in engine room, see figure 4.4.2. The StratCat, designed by BMT Nigel Gee, made a speed of 27 knots fully loaded at 130 T displacement and had a bollard pull of over 16 ton with the IPS pod drives. Six of those vessels was built for Njord Offshore in UK. The vessel with slender hulls and IPS900Quad with dynamic position system is shown in figure 4.4.3.

This type of vessel was followed by similar vessels with WJ drives and with CPP, but no other propulsion system could give same overall performance as the IPS900 could. The redundant system would give a speed of 27 knots on 4 engines and 21 knots with 3 engines. So should one engine go down, the crew will always get back to harbour in the evening. Out at the wind farm, the flexible engine solution will give the possibility to run the vessel on two engines only, by that saving fuel and engine running hours and increase time between overhaul.

Figure 4.4.2 IPS900 Quad engine room

Figure 4.4.3 NJORD 27m CTV with IPS900Quad and DP (picture from Njord Offshore)

As mentioned already in section 4.2, the vessels capability to carry cargo and not losing speed is an important parameter when judging the vessels overall performance. With high load carrying capability more passengers and cargo can be transferred out to the wind farms, SOVs or Installation vessels. In figure 4.4.4 the load carrying capability of IPS compared with WJ, FPP and CPP is shown, taken from a few examples of modern CTVs, one catamaran type and one mono hull type. Figures are taken from full load displacement condition to give a fair comparison. The trend lines are clear and show that the IPS system are on top of the other propulsion systems. Volvo Penta IPS will generate more payload per horsepower.

*** Formatting Issue - Best Available Paper/Graphic ***

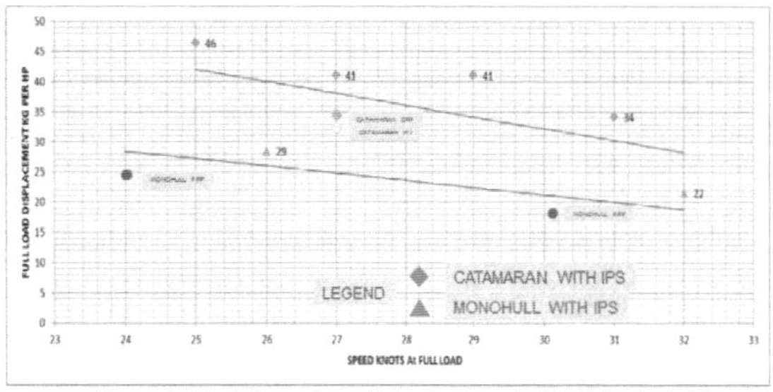

Figure 4.4.4 Load carrying capability

As previously discussed, the Volvo Penta IPS has proven to be very fuel efficient. The potential fuel savings in a 3000 hour per year offshore wind operation is approx. 20% based on a wind farm duty cycle. With an average fuel consumption of 200 litre per hour for the vessel with IPS900 Quad will give a fuel saving of more than 40 litre per hour compared to other type of propulsion systems. This will sum up to 120.000 litre savings per year equivalent to a CO_2 reduction of 318 ton compared to WJ or FPP solutions.

The NJORD Offshore vessels with IPS900Quad built by Strategic Marine in Singapore and now also by Cheoy Lee in China, have accumulated today over 230.000 engine running hours and counting. The savings of fuel compared to other propulsion systems, with an average of 50 litre per hour per engine will be around 2.3 million litres with the subsequent reduction in carbon footprint with 6095 ton over the past 3 years!

This you could say is just the beginning on the road to more sustainable wind farm services, with further actions to be described in next chapter.

5. FUTURE DEVELOPMENTS

5.1 EFFICIENT AND CLEAN PROPULSION

As a global supplier of marine engines, marine gensets and complete propulsion systems, Volvo Penta has taken further steps towards a more sustainable future. As described above, Volvo Penta is putting a lot of resources into the development of efficient propulsion systems. We believe that the biggest potential in fuel savings and CO_2 reductions (in %) lies in efficient propulsors, designed for certain speeds, such as Aquamatic stern drives and IPS pod drives. They both carry the DUOPROP technology for Contra Rotating propellers, which in all speeds, planing or displacement will create efficiency gains from 10% up to 35% compared to single screw. Such propulsors, coupled to combustion engines using fossil free fuel, like HVO (synthetic diesel) will reduce the CO_2 emissions in two steps. First by using less energy for a certain speed and second by using energy sources that are CO_2 neutral. In addition, the combustion engine should be equipped with a catalyst to reduce NOx and particulates when desired and when required.

Figure 5.1.1 Volvo Penta D13 with SCR catalyst

See above figure 5.1.1 showing a complete integrated propulsion package with engines, drives, shafts, exhaust and SCR catalyst, all from one supplier.

5.2 ELECTRIC PROPULSION

The next move for a more sustainable future is using the renewable energy coming from the offshore wind turbines, from wave power, from solar power or from hydro power. For that you need an electric motor. Volvo Penta belongs to the Volvo Group, were you already can see electric hybrid and full electric buses and refuse trucks being delivered to cities all over the world. The parallel hybrid concept shown below in figure 5.2.1, based on common parts from the Volvo group, will be seen in operation during 2021. For an offshore wind farm the hybrid concept will work very well. From shore to the park, you may have travelling distances up to 200Nm. Unfortunately, there is no chance to go full electric on this voyage if you would like to make a reasonable speed above 20 knots. The batteries or energy storage system of today are too heavy, absorbing all your payload and more on top. But as you will spend at least 50% of the time out at the park, at lower speeds, there are good options that you can charge batteries at sea, which will fit the electric hybrid solution very well. So all operations out at the windfarm can be performed by using the electric motor only. And should you for some reason run out of power from your energy storage system (ESS), then the combustion engine will always back you up, for example if additional bollard push is required to stay safe against a turbine landing in really bad weather.

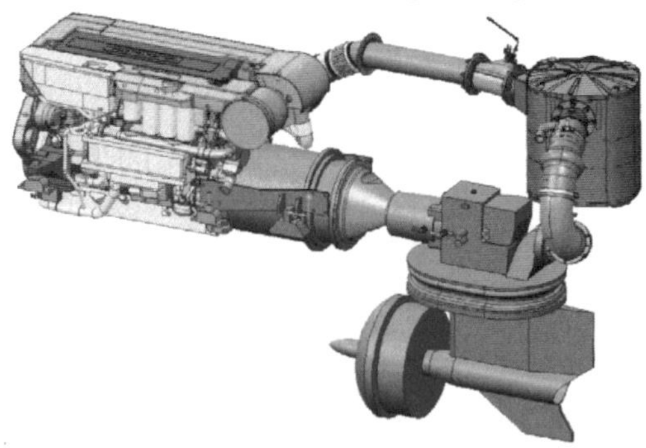

Figure 5.2.1 Electric hybrid propulsion package

It's clear that the windfarm developer needs to prepare for growing number of electric hybrid CTV projects already now. The available charging capacity at the turbine are normally less than 63 Amp, which is not enough for reasonable fast high-power charging. So, charging stations needs to be designed in to the offshore wind farm already at the project phase, being able to high power charging between 300-450 kW. It could be a purpose built charging station taking energy from the turbine or from a substation. It could also, in emergency, be a charging station at the SOV, with a larger ESS charged overnight by offshore wind power or using diesel generator set (from Volvo Penta driven by HVO, synthetic diesel) to generate electricity. The charging stations must of course be designed to sustain very rough weather conditions, sometimes ice cold, sometime very hot and humid and all time in a very corrosive environment.

Volvo Penta believes that within a couple of years, a majority of the new build support vessels will have some kind of hybrid propulsion system, in order to reduce CO_2 emissions. A number of vessels will also be retrofitted with hybrid systems, to save both fuel cost and CO_2, it goes hand in hand with the growth of the renewable energy sources and with rapid growth of electro mobility on and off road, in cities and in protected areas.
As a supplier of complete propulsion systems, Volvo Penta will offer full electric and electric hybrid systems, with all parts optimized to create the most cost efficient system, easy to install, easy to maintain, easy to run. The Volvo Penta system will include engines, motors, energy storage, switch boards, propulsors and drives, all designed to meet class requirements for offshore and people transportation.

6. CONCLUSIONS

The growth of SOVs has prompted development of smaller efficient CTVs. As the windfarms move further out from shore, the need and number of SOVs and hotel ships for the technicians positioned out at the park will grow as well. The smaller CTVs launched with davits and cranes provides the most flexible solution, making it possible for the CTVs to work parallel to gangway systems. This will maximize the daily operation efficiency and reduce OPEX. The more transfers per day you can perform, the more work will be done. Stern drives and pod drives are popular drive systems for those monohull vessels, keeping the smaller vessels in stable course and position during transit and transfer.

In the midrange CTVs, used for daily transit from safe harbours out to the wind farms, the catamarans will continue to be the main hull form. In order to increase the capacity and efficiency of the vessels you can now see the start of a number of retrofits coming up. It could be to install new propulsion systems to reduce fuel cost or introduction of hybrid systems to reduce emissions. Pod drives are very suitable for this type of retrofit, being compact with limited required space in engine room. In many cases engine room can be made shorter, giving room for an extra compartment for the energy storage system.

For the larger CTVs, normally using multiple engine installation for redundancy, the hybrid solution in combination with pod drives, will be a perfect match. For example, two of the drives could be with combustion engines and two of the drives could be hybrids with electric motors. During transit from harbour to the windfarm, the hybrid electric motors can be used for charging the batteries if needed and during transfer the electric motors can give an extra boost during bollard pull when required. While waiting or sailing around in the park, the electric motors on its own will be used to run the vessel between the turbines.

For a sustainable future, not only the engines needs to be CO_2 neutral, also the propulsions system should be designed in such way that the highest propulsive efficiency is achieved in desired speeds together with efficient hull forms. The contra rotating propellers used in the Volvo Penta sterndrives and IPS pod drives are definitely the best for achieving the highest propulsion efficiency over the whole speed range. The combination of efficient low weight combustion engines and electric motors together with slender pod drives is an unbeatable combination for achieving low fuel cost, low emissions and safe handling of the offshore crew and work boats, now and in the future.

7. ACKNOWLEDGEMENTS

Many thanks to Bjarne Mikkelsen,Senior Service manager at Esvagt, Robbert van Rijk, Managing Director of Windcat Workboats, Jonas Pedersen, Managing director of Tuco Marine, Andrew Thwaites, Technical Director at Njord Offshore, Martin Landström,Chief Business Development Officer at Northern Offshore Services and Henrik Myklebust, Sales Support at Maritime Partner for valuable inputs and technical discussions.

8. REFERENCES

1. LAWLESS, J, BMT Nigel Gee, '26m Crew Transfer Vessels & Their Development.', ,*Shaping Offshore energy solutions, Gothenburg, 26 June, 2015.*

2. TORNEMAN, G., 'Multiple Pod units for efficient vessel handling in wind farm operations', *RINA WFSV Conference, London, 29th January, 2015.*

3. https://www.carbontrust.com/media/674745/carbon-trust-p-plot-development-research-summary-june-2017.pdf

***** Formatting Issue - Best Available Paper/Graphic *****

4. https://www.carbontrust.com/media/674716/p-plot-seminar-presentation-june-2017.pdf

9. AUTHORS BIOGRAPHY

Gerard Törneman, M.Sc in Naval Architecture and Marine Engineering from Chalmers University of technology, holds the current position of Sales Project Manager at Volvo Penta in Sweden. He is responsible for business development within the Marine Commercial segment, particular in the offshore energy sector. His previous experience includes being General Manager at MJP Marine Jet Power in Sweden and Project Manager for MacGregor Scandinavia.

A COMPARATIVE ASSESSMENT OF CREW TRANSFER VESSEL MOTIONS BETWEEN INCLINED AND VERTICAL BOAT LANDING ARRANGEMENTS

O Hizir, M Maasch, O Turan and S Day, University of Strathclyde, UK
I Nicholls, Scottish and Southern Energy (SSE), UK

SUMMARY

Crew Transfer Vessels (CTVs) are an essential part of operations and maintenance activities for offshore wind farms. CTVs, which are generally multihull vessels transport crew from shore to offshore wind farm sites to carry out the scheduled maintenance and repair activities. CTVs thrust onto turbine structures from the rendered bow and push against the structure creating frictional thrust to prevent the bow from relative motions. This study presents and discusses the results from a set of model experiments and numerical calculations to allow a comparative assessment of a CTV's landing-manoeuvre performance between inclined and vertical boat landing arrangements. The hydrodynamic performance of a CTV in open water and the "thrust-in" condition using inclined and vertical landing arrangements are presented for different vessel thrust force and different fender properties.

NOMENCLATURE

CTV	Crew Transfer Vessel
OWT	Offshore Wind Turbine
O&M	Operations and Maintenance
RANS	Reynolds Averaged Navier-Stokes
BEM	Boundary Element Method
KHL	Kelvin Hydrodynamics Laboratory
BP	Bollard Push
μ	Fender Friction Coefficient
T	Fender Compression Reaction Force
Ff	Fender Friction Force
VL	Vertical Landing
IL	Inclined Landing
H	Regular Wave Height

1. INTRODUCTION

CTVs are the fast and cost-effective means of the transfer of maintenance personnel and equipment to the fixed OWTs. They are generally catamaran type vessels and their length range between 12 and 24m. Small vessels are operating in protected areas up to one meter of significant wave height whereas large ones are operating in more exposed areas with up to two meters of significant wave height [1]. According to [2], approximately 25% of the total cost of producing wind energy from offshore wind farms arises from their O&M. Following the current trends, OWTs are getting bigger and located in the open sea as further as their fixed foundations are feasible with the increasing water depth. This will increase the O&M costs and push the designers and operators to find more effective means of transportations to the OWTs.

The landing manoeuvre of a CTV to a fixed OWT foundation is a very complex and a multi-physics problem which involves the hydrodynamic interaction of a vessel and a turbine foundation and the mechanical reaction of the fenders under resultant forces which is the sum of the hydrodynamic and bollard push forces. Due to the complexity of the problem, researchers generally used a partitioned solution approach to solve hydrodynamic and mechanical problems separately and couple them accordingly to obtain a valid solution.

Recently, a couple of researchers investigated the landing manoeuvre and the operability of CTVs. [2] investigated the landing manoeuvre of a CTV to the turbine foundation using time-domain method implementing the slip and stick conditions to the main motion equations. The author also investigated the operability performance of a particular CTV vessel. [3] implemented a partitioned solution approach to solve multi-physics problem. The study used a mesh-based multiphase finite-volume RANS method, which facilitates an overset-grid technique to address large relative motion between multiple floating objects and a quaternion-based motion model, which constraints the bodies due to mechanical connections. Authors indicated that the combination of generic mechanical joints between multiple (floating) objects and overset grid methods are a viable option to analyse complex hydrodynamic scenarios at manageable computational costs if a short-time analysis is sufficient. [4] and [5] also used a partitioned solution method to solve the landing manoeuvre problem. Authors investigated the vessel and OWT foundation hydrodynamic and mechanic interaction problems at the landing instant using numerical methods and comparing them with the experimental results. The numerical hydrodynamic interaction problem is solved using potential theory using a BEM and the solved using a fender model, considering the reaction forces due to the friction and the deformation of the fender. Numerical simulations showed a good agreement with the experiments in slip condition, but underpredicted the stick condition responses due to the simplified fender models. The authors pronounced the importance of using non-linear fender models in the simulation whereas this results in a higher computational cost.

This study presents and discusses the results from a set of model experiments and numerical calculations of a CTV's landing-manoeuvre performance focusing on the influence of the VL and IL arrangements.

2. EXPERIMENTAL SETUP

2.1 EXPERIMENTAL SETUP INTRODUCTION

This section presents the experimental setup of the hydrodynamic performance of a CTV in open water and the "thrust-in" condition using VL and IL rail arrangements for different vessel thrust force and different fender properties. The objectives of the experimental tests were as follows:

- Measure the friction properties of the idealised fenders to characterise the frictional thrust that restricts the vertical motion at the bow of the CTV.
- Measure CTV's motion responses in head seas in open water without landing rail interaction.
- Assess the difference between VL and IL for different fender and thrust force scenarios.
- Provide data for comparison with numerical models and validate the numerical predictions

2.2 KHL

All tests were carried out in the tank of the KHL at the University of Strathclyde. The tank is 76.2m long, 4.57m wide and 2.50m deep and is equipped with a four paddle active-absorbing wavemaker made by Edinburgh Designs with a hinge depth of 1.50m. The wavemaker is installed in a movable watertight structure which allows the tank to be run with a range of water depths from 1.6 – 2.3 m. Regular waves of over 0.6m in height and irregular waves of over a significant wave height of 0.75m may be generated. A beach of total length 13m is installed. The beach has a reflection coefficient of less than 5% over the range of frequencies of interest for this study. The model was installed in the centre of the tank both longitudinally and transversely.

2.3 COORDINATE SYSTEMS

In the experimental data acquisition, two coordinate systems were used: a body coordinate system CB (Xgb,Ygb,Zgb), located at the centre of gravity (COG) of the CTV and a fender coordinate system CF(Xf,Yf,Zf), located at the centreline of the CTV and the centre of the fender. Figure 1 illustrates the origin points of the body coordinate system and the fender coordinate system with respect to the forward and the aft perpendicular.

2.4 BALLASTING AND MASS PROPERTIES

The tests were performed at a 1:10 scale, which allowed the use of an existing CTV model. The principal dimensions of the CTV in full-scale and in model scale are given in Table 1.

Prior to the seakeeping/boat landing experiments, the CTV model was prepared by weighing, inclining and swinging according to [6]. The fully equipped CTV model was ballasted until the required model displacement was reached. An inclining test was then carried out to determine the vertical distance from the keel (lowest point of the keel line) to the vertical centre of gravity (KG), and the vertical location of ballast weights adjusted to set the desired value of KG. Finally, a "swing" test was used to determine the pitch radius of gyration (K_{YY}) of the CTV model by measuring the period of oscillation, and the weights moved longitudinally to set the desired target value of the radius of gyration. This process ensured that the correct mass properties of the CTV model were set since these may strongly influence the seakeeping performance.

Figure 1: CTV coordinate systems

Dimensions	Full Scale	Model Scale	Units
LBP	18	1.8	m
B	6.9	0.69	m
T(design)	1.38	0.138	m
Tf	1.38	0.138	m
Ta	1.38	0.138	m
Displacement	45.4	44.293	tonnes/kg
Volume	44.293	0.044293	m^3
Kg	2.557	255.7	mm
KM	10.55	1055	mm

Table 1: Full scale and model scale principal dimensions

Majority of catamarans have light superstructures compared to other hull designs. It is a general practice to assume the roll radius of gyration of catamarans as Kxx=0.4*B. A literature review is performed for the similar size catamaran designs and it has been observed that [7] calculated the roll gyradius is assumed to be Kxx=0.413*B whereas [8] calculated the roll gyradius as Kxx=0.443*B. In order to be consistent with the former studies, in the present experiments roll radius of gyration is assumed to be Kxx=0.4*B.

Results	Full Scale	Units	Model Scale	Units
GM measured	7.682	m	768.205	mm
KG measured	2.62	m	262	mm
KG corrected	2.575	m	257.538	mm
GM corrected	7.727	m	772.7	mm
Abs KG Error	-0.0184	m	-1.84	mm
% KG Error	-0.719	%	-	-
Kyy/LBP	0.295	-	0.295	-

Table 2: Full scale and model scale principal dimensions

2.5 TANK TESTS SETUP

All seakeeping tests were carried out in head seas with a tank water depth of 2.0 m. The CTV motion response analysis in open water uses the body coordinate system CB as the motion reference point. This is the standard convention for ship motion measurement. However, for the landing manoeuvre, all data is presented in the fender coordinate system CF to allow easy evaluation of the motion response at the point of contact with the boat landing.
The CTV motion responses were investigated to determine the hydrodynamic performance characteristics in regular head waves at zero forward speed, representing idealised operational conditions.
The waves in the tank were measured using standard resistance wave probes. Vessel motions were measured in six degrees of freedom using a Qualisys optical motion capture system. The markers for the Qualisys system can be seen as reflective balls in Figure 2.

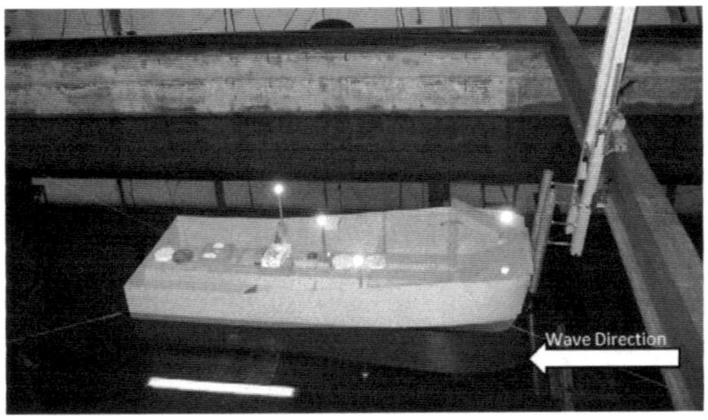

Figure 2: Experimental setup and wave direction

The jacket was not modelled in these tests. While the presence of the jacket may well influence the incoming waves in the real condition, it was agreed that a comparative study would neglect this effect of the jacket and focus on the inclination of the boat landing. It is expected that this is likely to be a conservative assumption, since the presence of the jacket upwave of the boat landing may well reduce the wave height at the boat landing.
The thrust-in force was simulated using a wire attached to the bow of the vessel. This was led horizontally around 10m upwave and over a pulley. A weight hanger was suspended from the wire; thus the weight placed on the weight hanger simulated the thrust force. The relatively large distance of the pulley upwave of the boat meant that even when the bow of the boat was moving vertically, the change in angle of the wire was extremely small, and the influence of the wire on the vertical motion could be assumed to be negligible.

2.6 FENDER FRICTION TESTS

For the accurate experimental testing of a landing-manoeuvre performance of a CTV, the frictional properties of the deformable fender material are required. However, information on the properties of the fenders to be used in this project was not available. Therefore, several different rubber materials were tested in the KHL. The experimental setup is shown schematically in Figure 3 at a scale of 1:10, consisted of the boat landing bars represented by two steel pipes (green) and a fender represented by two rubber tubes mounted on a base frame (red). The fender was weighted in its centre by varying masses, which simulated a varying thrust force. Over a pull cord attached to the fender, another set of varying masses simulated the applied forces.

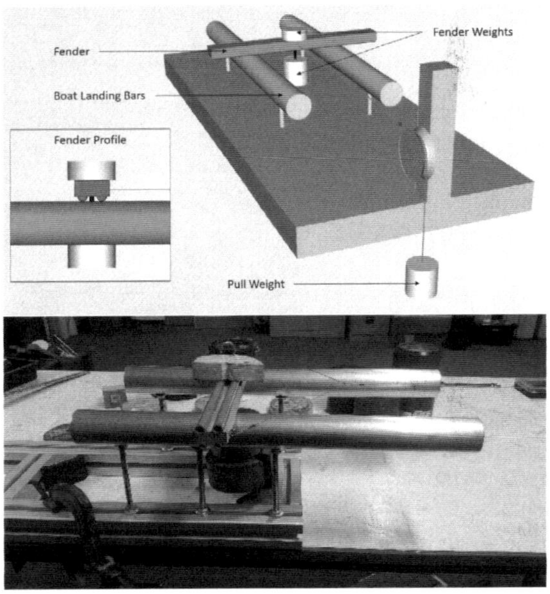

Figure 3: Experimental setup for friction tests

Three rubber materials, Material I, and Material II, Material III were chosen. For different fender weights, the pulling weight was gradually increased until instant the fender would start to slide along the boat landing bars. This allowed estimating the minimum threshold force that was required to move the fender. This approach refers to the Coulomb friction where the static friction coefficient was assumed to equal to the kinetic friction coefficient (See Figure 4).

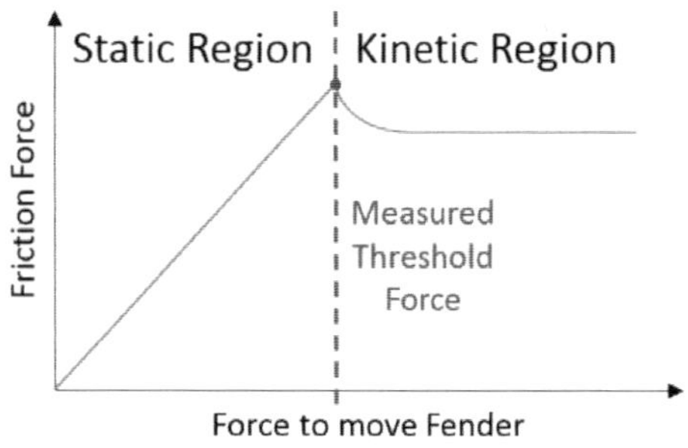

Figure 4: Static and kinetic friction

Stiction is the static friction that needs to be overcome to enable relative motion of stationary objects in contact. The stiction friction coefficient was calculated using the minimum threshold force needed to prevent relative motions between two surfaces. Material II was tested in dry and wet conditions whereas Material I and Material III were only tested in dry conditions. It can be observed from Figure 5 that Material II yielded stiction coefficients of $\mu_{II-dry} = 0.12$ and $\mu_{II-wet} = 0.07$ whereas Material I and Material III yielded $\mu_{I-dry} = 0.35$ and $\mu_{III-dry} = 0.40$ respectively.

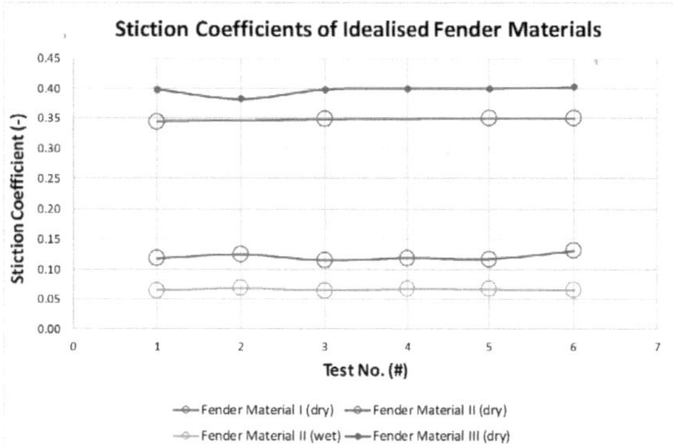

Figure 5: Stiction coefficients for different fender materials

In the following sections, the fender with a stiction coefficient of μ=0.12 will be named as "stiff fender", the fender with a stiction coefficient of μ=0.35 will be named as "soft fender".

It should be noted that the landing rails were manufactured from aluminium tube, and were extremely smooth. Hence the friction coefficients obtained may be rather lower than those which might be found in an operational environment. Hence the results of the tests may over-estimate the full-scale values. However, the comparison between the two landing cases is still expected to be correct.

It is important to point out the influence of the VL and IL arrangement to the friction coefficient and the maximum friction threshold force. Friction between interacting surfaces is a physical property that depends on the wetness, roughness, temperature and humidity between the interacting surfaces. Therefore, once the aforementioned physical properties are identical between the interacting surfaces, the orientation of the landing arrangement does not affect the value of the stiction coefficient.

3. NUMERICAL ANALYSIS

3.1 NUMERICAL ANALYSIS BASIS

This section presents the basis of the numerical calculation of the hydrodynamic landing-manoeuvre performance of a CTV. Numerical calculations were performed in open water and the "thrust-in" condition using VL and IL rail arrangements for different vessel thrust force and different fender properties. The objectives of the numerical calculations were as follows:

- To mimic the experimental friction properties of the idealised fenders in numerical calculations to characterise the frictional thrust that restricts the vertical motion at the bow of the CTV.
- To calculate CTV's motion responses in head seas in open water without landing rail interaction and by using the critical damping of motion modes match the motion RAOs with corresponding the experimental test results.
- To validate the difference between VL and IL for different fender and thrust force scenarios.

The CTV motion responses were investigated to determine the hydrodynamic performance characteristics in regular head waves at zero forward speed, representing idealised operational conditions. The thrust-in force was simulated using the bow of the vessel where a wire was attached in the experiments. In numerical calculations thrust force direction is constant and not depending on the CTV motions.

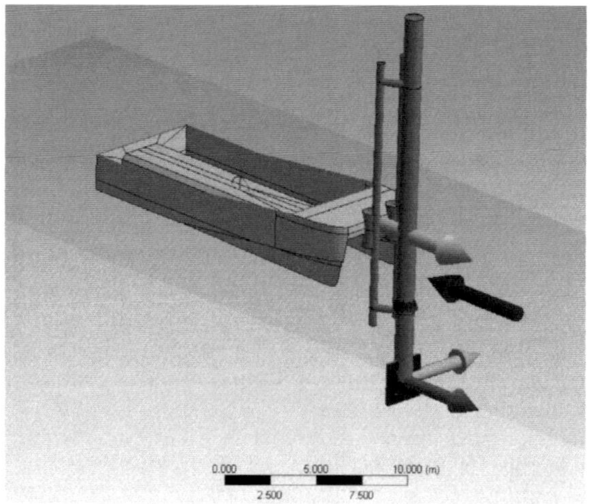

Figure 7: Numerical simulations setup and wave direction (Arrows: Yellow – BP direction, Blue – Wave Direction)

3.2 ANALYSIS TOOLS

The hydrodynamic analysis of the CTV landing manoeuvre is performed using the commercial software ANSYS 19.0 hydrodynamics module AQWA. The AQWA software is a modularised, fully integrated hydrodynamic analysis module based on a 3-D diffraction and radiation solver, in addition to Morison equations. The AQWA module has the capability of modelling multiple interacting geometries, mooring lines, articulations and fenders and can transfer motions and loads to the dedicated finite element analysis (FEA) modules.

3.3 NUMERICAL ANALYSIS COORDINATE SYSTEMS

In the numerical analyses, identical coordinate systems are used with the experimental tests to maintain the consistency. Figure 7 illustrates the origin points of the body coordinate system and the fender coordinate system.

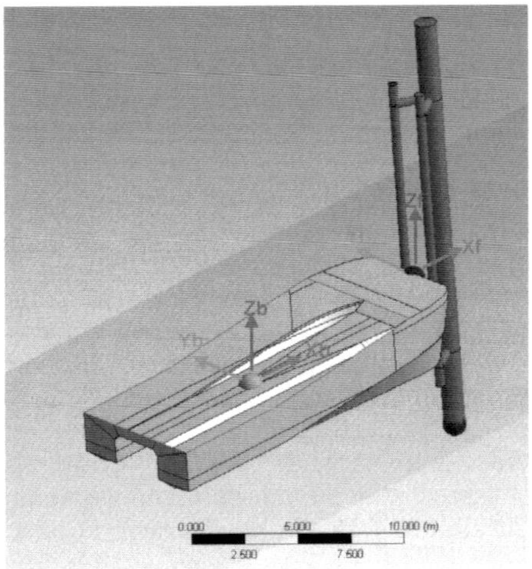

Figure 7: Fender and body coordinate axes in numerical analyses

3.4 EQUATIONS OF MOTION

In the simulations, the combined low frequency and wave frequency motions of each body are calculated in 6 degrees of freedom in the frequency domain and time domain. In the time domain, simulations are performed using an impulse-response functions (IRF) approach, and the degree of freedom of the system can be reduced using articulations. The linear time-domain equation of motion is defined using the IRF as shown below:

$$[M_{jk}^h + A_{jk}^h(\infty)] \cdot \ddot{\xi}_k(t) + \int_0^t K_{jk}^h(t-\tau) \cdot \dot{\xi}_k(\tau) d\tau + C_{jk}^h \cdot \xi_k(t) = F_j^I(t) + F_j^D(t) \quad \text{j,k=1,2...6} \quad (1)$$

where the subscripts j,k stand for the motion modes. The left side of the equation gives the fluid reaction forces and inertial forces while the right-hand side of the equation gives the excitation forces in the time domain. In the equation, M_{jk}^h is the mass and inertia matrix of the floater, ξ_k , $\dot{\xi}_k$ and $\ddot{\xi}_k$ are the time-domain displacement, velocity and acceleration vectors respectively, $A_{jk}^h(\infty)$ is the infinite frequency added mass, K_{jk}^h are the memory functions for related motion modes, C_{jk}^h is the constant restoring force matrix, $F_j^I(t)$ and $F_j^D(t)$ are the incident and diffraction forces in the time domain. The present study focuses on the nonlinear derivation of motion and load responses therefore the linear restoring matrix is not used; instead, restoring forces are calculated by the direct pressure integration method using the exact wetted area of the floater at each time step in the simulations. The memory functions K, the added mass coefficients A and diffraction forces are determined using the results of the radiation-diffraction boundary value problem solution.

3.4 IMPLEMENTATION OF FENDERS TO THE NUMERICAL SIMULATIONS

In AQWA, elastic fenders need to be implemented in the motion equations in order to mimic the elastic behaviour of fenders and the slip/stick conditions during the landing manoeuvre of the CTV. In AQWA, fenders are the only way to model contact between two structures, so if there are no fenders, structures can pass through each other. In the present study two types of fenders are used (see Figure 8 and 9) to simulate the contact between two structures:

- Fixed Omni-Directional fender, which is connected to one of the bodies and contact with the second body may be anywhere on the surface of the fender (assumed spherical). This type of fender was used to model the bow fender of the CTV interacting with the landing rails.
- Fixed Directional fender, which is connected to one of the bodies and contact with the second body is assumed to act in a given direction. This type of fender was used to model additional friction and stiction forces acting on the CTV bow section during the landing manoeuvre.

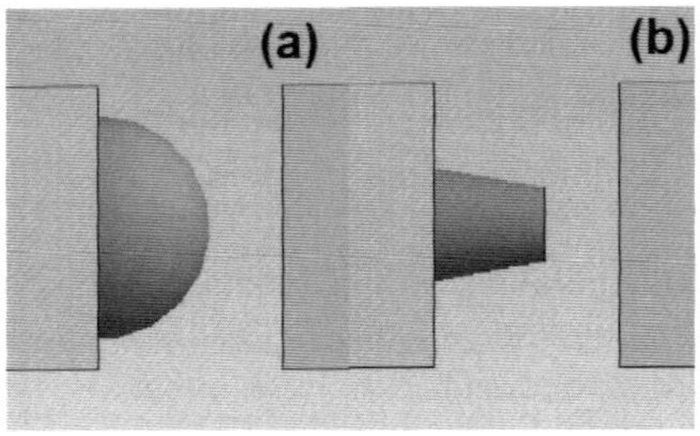

Figure 8: Omni-directional (a) and fixed directional fender types.

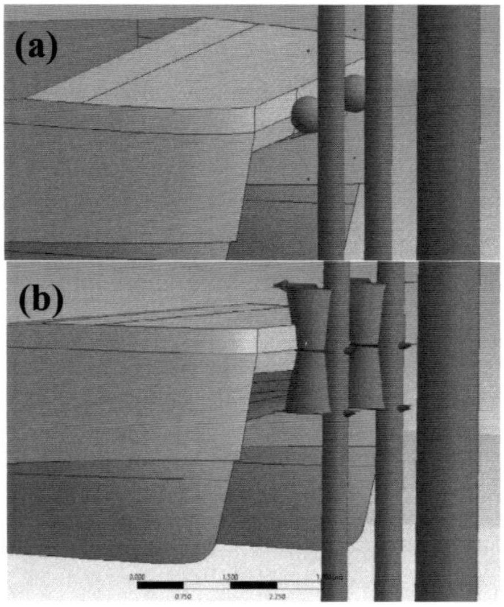

Figure 9: Omni-directional (a) and uni-directional fender locations.

3.6 FENDER FRICTION AND DAMPING

The fender friction force on the contacting surface is given by $F_f = \mu * T$ where μ is the friction coefficient, and T is the normal compression reaction force. In the CTV landing maneouvre simulations the friction coefficients were taken from the physical friction experiment results. In AQWA fender friction works best in situations where the friction force is smaller than other forces in the same direction. The friction will slow down relative motion between two structures, but is not suitable for keeping them fixed together (no stiction). When the relative velocity changes sign the friction force must also change sign, but to avoid an instantaneous change in this force (and therefore an instantaneous change in acceleration) a smoothing function is applied. This means that when the relative velocity is very small the friction force is also small, which may allow some relative motion between the two structures.

A fender damping force is only applied in the direction of the fender compression force and is modelled by a linear material damping, as:

$$F_d = \beta K_f \frac{dL_d}{d_t} \tag{2}$$

where β is the damping coefficient and K_f is the fender stiffness. In the CTV landing analyses, additional friction force acting on the bow fender of the CTV is applied using the fender damping force method. In reality, the friction force can only be applied using the damping force, otherwise if the stiffness coefficient were used additional fenders would push back the CTV to its original position which would not mimic the real behaviour. The disadvantage of this method is damping is a function of fender compression velocity and applying force through damping translates the motion signals without changing its amplitude. In order to generate accurate friction forces, four additional fenders are defined to act opposite way of the CTV motion (see Figure 9b).

4. NUMERICAL ANALYSIS VALIDATIONS

This section presents the validation of AQWA numerical calculation results of the landing manoeuvre of the CTV with the experimental test results. This section investigates thoroughly the forces applied to the fenders during the simulations. The investigated fender properties during the landing manoeuvre are summarised below:

- Fender Compression
- Total Fender force applied by the CTV
- Fender friction force
- Additional fenders damping force

All numerical tests were carried out in head seas, at a peak response period of T=5.2s and at a wave height of H=0.4m to compare the resultant vertical bow motions with the available experimental results.
As the material properties of the full-scale fender and boat landing platform were unknown, the stiffness and damping coefficients of the idealised deformable fender materials calculated to mimic the same landing manoeuvre behaviours of the CTV whereas the frictional properties were determined prior the tank tests.

The omnidirectional fenders' damping and stiffness coefficients have a significant influence on the landing simulations. The stiffness coefficient of the omni-directional fenders needs to be small enough to prevent CTV bounce back from the landing rails due to the elastic force. Moreover, the damping coefficient of the omni-directional fenders needs to be high in order to damp the motions and keep the CTV stable in its landing course. In order to satisfy these behaviours and mimic the behaviour in the experimental tests, the omni-directional fenders properties were designed as shown in Table 3.

Omni-directional Fenders Physical Properties		
Damping Coefficient (kNs/m)	Stiffness Coefficient (kN/m)	Initial Size (cm)
320	800	30

Table 3: Omni-directional fenders physical properties

In numerical simulations, the maximum elastic compression of the omni-directional bow fender was observed as 5cm at the aforementioned wave conditions and at 7 tons of BP.

It was mentioned before that the fender friction force is a function of compression force T and the friction coefficient. The omnidirectional fenders were tested at a peak period of 5.2s and H=0.4m and 7 tons of BP applied. The omnidirectional fender compression force was extracted from the simulations and given in Figure 10. It can be observed from the figure that the total fender compression force, which is the resultant force from stiffness and damping force components, is higher in the IL arrangement compared to the VL arrangement.

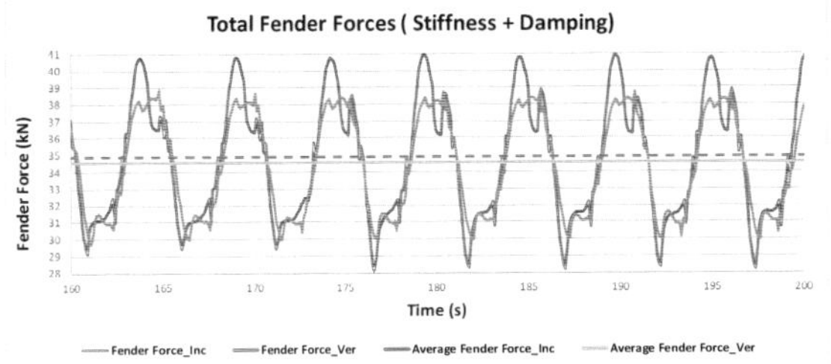

Figure 10: Omni-directional fender compression force in VL and IL arrangements.

In AQWA a smoothing function is applied when the relative motion between structures are small. The smoothing function in the friction calculations sometimes causes incident friction drops in the simulations. A VL simulation using stiff fender (μ=0.12) with identical wave conditions and BP was tested and it can be observed in Figure 11 that the smoothing function caused instant friction force drops in the calculations. It can also be observed that the calculated friction forces and the expected friction force which is $F_f = \mu * T$ agrees well except the instant friction force drops and the average expected friction force is slighly higher than the calculated average friction force. It is also beneficial to emphasise that figures are only presented for the port-side fender due to the symmetrical forces applied to both port and starboard fenders in head seas. However, in other wave directions, the compression on the port-side and starboard fenders may be different, hence this may results in imbalanced friction forces.

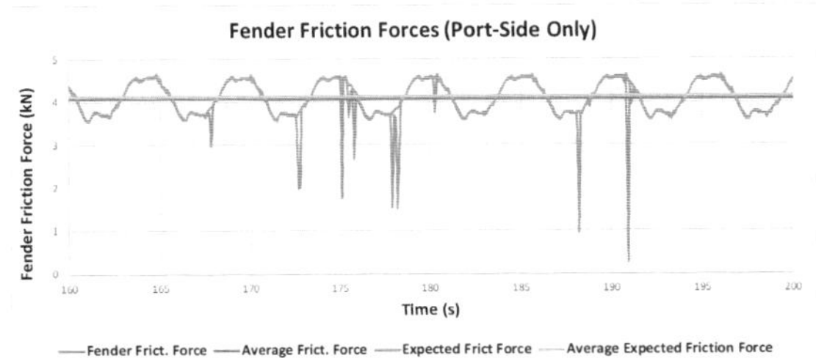

Figure 11: Omni-directional fender friction force in VL arrangement

After tedious test to eliminate the instant friction drops and match the vertical bow motions in the experimental tests, the uni-directional fender properties were defined in the numerical calculations where their physical properties are given for the stiff fender in Table 4.

Uni-directional Fenders Physical Properties		
Damping Coefficient (kNs/m)	Stiffness Coefficient (kN/m)	Initial Size (cm)
10	1	100

Table 4: Uni-directional fenders physical properties for stiff fender

Finally, the numerical and experimental comparison of the relative bow vertical displacements were performed at the peak wave period of 5.2s and wave height of 0.4m as illustrated in Figure 12.

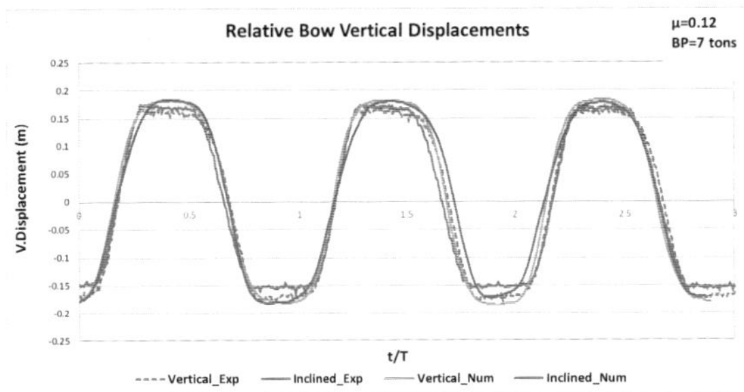

Figure 12: Comparison of Numerical and Experimental vertical displacement time in VL and IL at fender location at H=0.4m and T=5.2s (BP=7 tons, Stiff fender)

Following results can be drawn from the numerical and experimental validation case:

- In both VL and IL arrangements numerical simulations follow the experimental trend well.
- In both landing arrangements, numerical results are slightly higher compared to the experimental test results.
- IL arrangement bow vertical motions are smaller compared to the VL bow vertical motions. This is due to higher compression force occurred in the IL simulations compared to the VL simulations.
- The shift in the bow vertical motion responses is due to the additional fenders in which damping force is the main source force applied in the numerical calculations.

5. NUMERICAL ANALYSIS AND EXPERIMENTAL TESTS RESULTS & DISCUSSION

5.1 CTV MOTION RESPONSES IN OPEN WATER

The numerical CTV motion responses were investigated with no landing rails present. The presented motion RAOs were calculated at the body coordinate system CB located at the centre of gravity. The additional critical damping coefficient (see Table 5) were calculated to match the experimental test results values in the numerical calculations.

Additional Critical Damping (Frequency Independent)		
Heave	Roll	Pitch
20 kN/(m/s)	30kNm/(°/s)	10 kNm/(°/s)

Table 5: Additional Critical Damping Coefficients

The comparison between the numerical and experimental results of the CTV motion RAOs in head seas are given in

Figure 13, for the heave, pitch and surge RAOs respectively.

Following conclusions can be drawn from open water motion responses of the CTV:

- The frequency for the peak response in the heave RAOs is observed around 8s wave period which corresponds to λ/LBP=5.41
- The frequency for the peak pitch responses is observed around 4.5s wave period which corresponds to λ/LBP=1.75.
- It can be seen from the RAOs that the pitch motion couples into both heave and surge since there are local peaks in the heave and surge RAOs at the peak frequency of pitch response.
- In numerical calculations without additional critical damping coefficients applied to the system, pitch RAOs are over-estimated around the natural period.

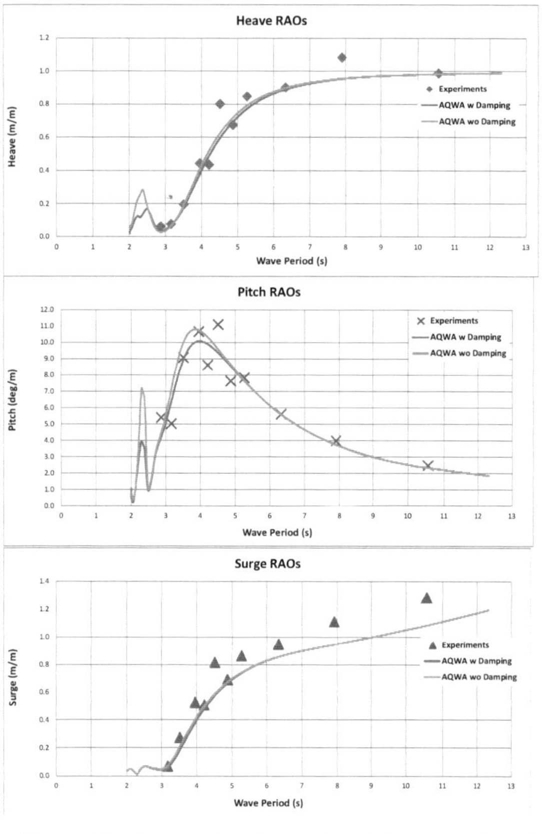

Figure 13: Numerical and experimental tests results of the heave, pitch and surge RAOs in head seas (Full Scale)

5.2 CTV LANDING MANOEUVRE MOTION RESPONSES BASIS

In this section, the numerical CTV landing motion responses are compared with the experimental results and discussed in detail. For the accurate representation of the data, all motion responses are given with respect to the fender coordinate system CF in landing simulations. This allows analysing the motion directly at the point of contact between CTV and landing rails. In particular, it should be noted that the heave motions presented correspond to the vertical displacements calculated at the bow of the CTV at the fender coordinate system CF. The characterisation of the CTV motions when the fender is in contact with the landing structure is not straightforward since the motions are influenced by some factors:

- The influence of the vertical and inclined landing arrangements.
- The influence of the thrust force.
- The influence of the fender friction coefficient.

5.3 CTV LANDING MANOEUVRE MOTION RESPONSES IN VERTICAL AND INCLINED LANDING ARRANGEMENTS

This section investigates the differences in the motion responses between vertical and inclined landing scenarios using numerical and experimental results. All numerical simulations were performed with identical experimental conditions of a H=0.4m and constant full-scale thrust force of 7 tons and 9tons (full-scale) using the tested stiff (μ=0.12) and soft (μ=0.35) fenders.

Figure 13 presents numerical and experimental results of the vertical bow motion RAOs comparison for the VL and IL using the stiff fender at 7 and 9 tons constant BP force. Fender physical properties which were given in Table 3 and 4 were applied in the stiff fender numerical simulations.

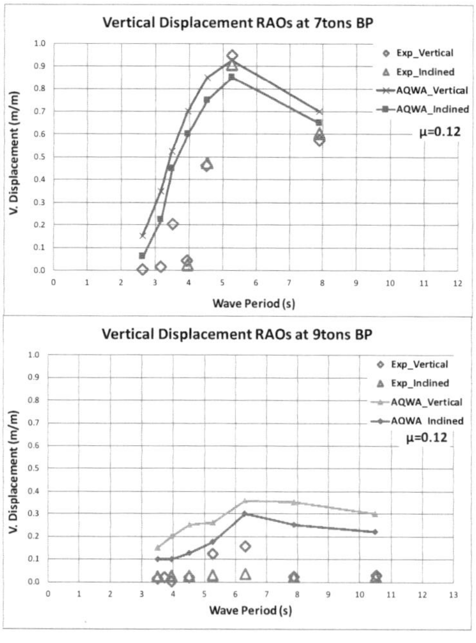

Figure 13: Vertical Bow Motion RAOs for VL and IL (BP=7 and 9 tons, Stiff fender)

It can be seen that the bow motions are quite substantial when the CTV has a constant 7 tons of BP and the stiff fender is used, suggesting that the fender friction coefficient is too low to restrain the CTV successfully even in the small waves used. The largest motions occur at frequencies close to the peak of the heave RAOs. Nonetheless, it may be seen that the numerical simulations trend agrees well with the experimental test trend whereas the results of the inclined landing show smaller vertical displacements at the peak frequencies than those for the vertical landing in both numerical simulations and experimental tests. It can also be observed that numerical simulation results are higher compared to the experimental results. This can be justified as the lack of viscosity forces and stiction in AQWA numerical simulations which results in an increase in the motion responses compared to experimental results.

The corresponding results for the 9-ton BP case show significantly smaller vertical displacement RAOs than the tests with 7 tons BP case. It is suggested that the reduced vertical displacement RAOs originate from increased stiction force due to the higher thrust force, which acts as a constraint on the vertical motions. It can be seen that the bow motion is virtually eliminated in the experiments with the inclined landing and that the motions are slightly larger with the vertical landing than with the inclined landing. Moreover, it is observed that numerical simulation results followed the experimental test trend, but due to the lack of stiction forces in AQWA numerical results show higher vertical displacements for both vertical and inclined landing arrangements compared to experimental tests. Nonetheless, it may be seen from the numerical simulations that results of the inclined landing show smaller vertical displacements at the peak frequencies than those for the vertical landing.

A corresponding set of numerical simulations were carried out with the soft fender at 7 and 9 tons of constant thrust force. Figure 14 presents the motion RAOs comparison for the inclined and vertical boat landing.

In the soft fender (μ=0.35) numerical simulations additional fenders were used with the following physical properties:

Uni-directional Fenders Physical Properties		
Damping Coefficient (kNs/m)	Stiffness Coefficient (kN/m)	Initial Size (cm)
200	1	100

Table 6: Uni-directional fenders physical properties for the soft fender

It can be seen that in this case, the bow motion is effectively suppressed for both landing conditions over all the wave frequencies tested, and hence no meaningful difference is observed between the two cases in the experimental tests. It is expected that the higher deformation in the soft fender generated a larger contact area between the fender and the landing rails and hence

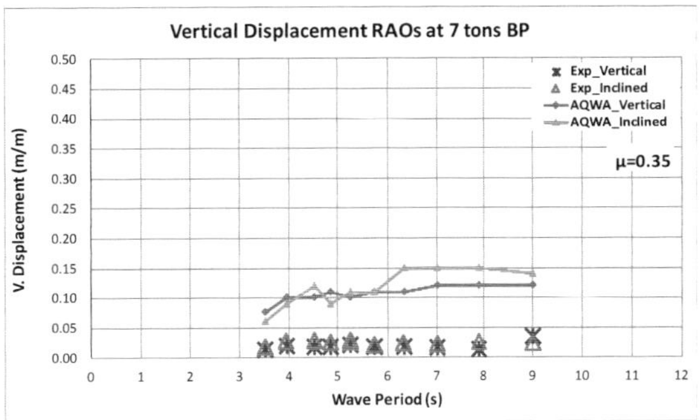

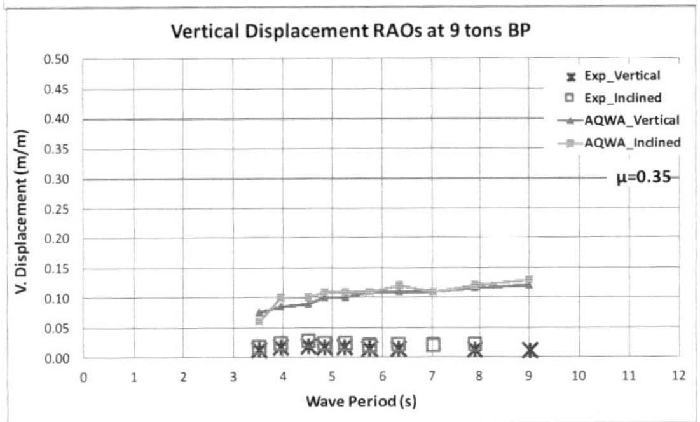

Figure 14: Vertical Bow Motion RAOs comparison of the VL and IL (BP=7 and 9 tons, soft fender)

higher friction forces applied to the CTV. The motion was effectively suppressed in all cases with 7 and 9 tons thrust, and it is suggested that at H=0.4m over all the wave frequencies the stiction forces are higher than the resultant tangential forces on the fender. In most of the cases resultant tangential forces it is observed that thrust force did not influence the results significantly for the tested wave height. Moreover, it can be seen that numerical simulation results trend agreed well with the experimental results trend where at 7 tons BP and around the peak period, the relative vertical bow motions are slightly smaller in inclined landing arrangement compared to the vertical landing arrangement. It is observed that due to the lack of stiction forces and viscosity forces numerical results show higher vertical displacements for both VL and IL arrangements compared to experimental tests.

6. CONCLUSIONS

This report presents the comparative assessment of CTV motions between inclined and vertical boat landing using various thrust forces and fender types (varying in frictional coefficient) obtained from numerical simulations and experimental tests. Following conclusions can be drawn from the results of the numerical simulations:

- In the majority of the cases, regardless of the thrust force magnitude, the CTV bow vertical motions are reduced in the inclined landing tests compared to the vertical landing tests. This is validated as in the IL arrangement fender compression force is higher compared to VL arrangement and this results in higher friction forces.

- When the full-scale thrust is increased to 9 tons, the vertical displacement RAOs are small for the stiffer fender. Nonetheless, vertical motions were still slightly larger for the vertical landing case.
- As expected, the soft fender, which has a higher friction coefficient, generally performed better in landing-manoeuvres compared to the stiff fender.
- In both landing arrangements, for a full-scale thrust force of 7 tons with the soft fender, bow vertical displacement RAOs are significantly smaller than those for the stiffer fenders. Motions are slightly larger for the vertical landing than for the inclined landing.

7. REFERENCES

1. Phillips, S., Shin, IB., Armstrong, C. 'Crew transfer vessel performance evaluation', Design & Operation of Wind Farm Support Vessels, 2015.

2. Sneep, T.P. 'Operability analysis of wind farm maintenance', Delft University of Technology Master Thesis, 2016.

3. Luo-Theilen, X., Rung, T., 'Computation of mechanically coupled bodies in a seaway', Ship Technology Research (Schiffstechnik), 2017.

4. Konig, M., Gonzalez, D.F., Abdel-Maksoud, M., Duster, A., 'Numerical investigation of the landing manoeuvre of a crew transfer vessel to an offshore wind turbine', Ships and Offshore Structures, 2017.

5. Gonzalez, D.F., Lemmerhirt, M., Abdel-Maksoud, M., Konig, M. ,Duster, A., 'Numerical and experimental investigation regarding the landing manoeuvre transfer vessel to an offshore wind turbine in waves', International Conference on Ocean, Offshore and Arctic Engineering (OMAE), 2015

6. International Towing Tank Conference guidelines for Seakeeping Experiments (ITTC 7.5-02.1), 2017

7. Van 't Veer, A.P., 'Behaviour of catamarans in waves', PhD Thesis, Delft University of Technology, 1998.

8 Soares, G., Fonseca, N., Santos, P., Maron, A., 'Model tests of the motions of a catamaran hull in waves', CEHIPAR, 2005

8. AUTHORS BIOGRAPHY

Dr Olgun Hizir is a naval architect and consultant at University of Strathclyde. He is working on offshore engineering fluid-structure interaction problems. He is responsible for R&D for seakeeping and hydrodynamic efficiency of ships.

Matthias Maasch is a PhD write-up student at University of Strathclyde. His main research focus is prediction of vessel hydrodynamics and hydrodynamic efficiency of vessels using Computational Fluid Dynamics methods.

Prof Osman Turan is a Professor in ship design, ship recycling and human factors at University of Strathclyde. He is responsible for teaching and research in ship design and marine hydrodynamics. He has extensive experience of various national and international ship design projects. He is directing the Department Research Excellence Framework (REF).

Prof Sandy Day is a Professor of marine hydrodynamics at University of Strathclyde. He is the Director of the KHL His research interests lie in computational and experimental hydrodynamics; he is actively engaged in fundamental and industry-focussed research on energy efficiency of ships, motions and loads on ships and offshore structures, and offshore renewable energy devices.

Ian Nicholls is a Naval Architect and Project Manager in Scottish and Southern Energy (SSE). He is responsible of the operations and maintenance of Beatrice Offshore Wind Farm. He has extensive experience in offshore wind farms maintenance planning and operation.

MICROBUBBLE – A UNIQUE TECHNOLOGY FOR REDUCING THE DRAG OF WIND FARM SERVICE VESSELS

Wenxian Yang and **Hongwei Yang,** School of Engineering, Newcastle University, UK
Ji Wang, Mingyang Smart Energy Group Ltd, Shanghai 200233, China
Meiping Sheng and **Minqing Wang,** School of Marine Science and Technology, Northwestern Polytechnical University, China

SUMMARY

Many experimental tests and model simulation research have been conducted previously in order to explore the potential of microbubbles in reducing the drag of vessels. However, different conclusions are drawn in different open literatures. In view of this, model simulation research is further conducted in this paper in order to seek the consensus on the influences of microbubble size, microbubble density, and microbubble initial velocity on drag reduction. From the model simulation results, it has been found that increasing microbubble density and their initial velocity will increase the local bubble volume fraction, thus increases drag reduction. However, the change of microbubble size will not influence drag reduction significantly. This research finding will be of very importance to the future design of low resistance wind farm service vessels.

NOMENCLATURE

ρ	Density of water (kg m^{-3})
μ	Dynamic viscosity of water (N s m^{-2})
μ_τ	Shear velocity (m s^{-1})
y_b	Distance from point *b* to the wall (m)
Y^+	Dimensionless wall distance
U_∞	The velocity of main flow (m s^{-1})
D_0	Drag force before applying microbubbles (N)
D_1	Drag force after applying microbubbles (N)
r	Drag reduction ratio
ϵ	Density parameter of microbubbles
v	Initial velocity of microbubbles (m s^{-1})
d	Microbubble diameter (m)

1. INTRODUCTION

Statistics has shown that approximately half of the ship's propulsion power is used to overcome friction resistance [1], which means that ship resistance is one of the main sources of fuel consumption of a vessel. This issue can be also observed from wind farm service vessels (WFSVs). At present, most WFSVs are catamarans with specific Small Water Plane Area Twin Hull (SWATH) design. The SWATH has small water plane area, which enables the motions of the vessel to be less affected by waves. In addition, the mutual balance effect of the twin hulls and the air tunnel between the two hulls can also benefit the stability of a SWATH vessel in voyage [2]. These advantages account for the popular application of SWATH to the design of WFSVs. However, a SWATH type WFSV uses two hulls to provide buoyancy. The large wetted surface area of the two hulls will inevitably lead to significant skin friction, which makes the fuel consumption of a SWATH type vessel much more than that of single hull variety vessels. It is reported that the energy required to be spent on a SWATH type vessel is 80% more than the energy spent on a mono-hull vessel of the similar size [3]. Therefore, the high fuel consumption of WFSV has been recognised as one of the critical issues in lowering the operation & maintenance cost of offshore wind farm. This issue must be addressed urgently, especially in the light of the fact that the offshore wind industry is rapidly growing and offshore wind farms are moving farther offshore in recent years [4].

Basically, the resistance of a vessel originates from three sources, i.e. skin friction, viscous resistance [5], and wave resistance. Among them, skin friction and viscous resistance, especially the former contributes a big portion of drag. It was reported that skin friction accounts for 60-70% of the total drag for cargo ships and contributes approximately 80% of the total drag for tanker vessels [6-8] in still water. This motivates the research of various forms of drag reduction technology in recent years. A comprehensive review of the relevant achievements can be found from [9]. Among the achieved technologies, one of the most attractive is microbubble technology [10] as this kind of technology is environmentally friendly, cost-effective, and moreover does not require to make any change to the hull form of the vessel. In order to explore this technology, many in-depth researches have been conducted in the past years by using either model simulation or experimental testing approaches [11-18]. However, up to date it is still not completely clear how microbubbles reduce vessel drag, although some believe that it is related to the variation of the viscosity of the fluid around the vessel or the modification of the turbulent structure in boundary layer on the hull surface of the vessel [19, 20]. In the meantime, microbubble production methods are also researched. The achieved methods include electrolysis [21], porous medium [22], venturi tube type bubble generator [23], and so forth. A comprehensive review of these methods can be found from [24].

Despite the aforementioned effort, many technical issues about microbubble technology are still unclear today. For example, a consensus on the influence of microbubble size on drag reduction has not been reached nowadays. Some said smaller microbubbles are more effective in reducing drag [25]; but others said that the variation of microbubble size did not have significant influence on drag reduction [26]. Therefore, it is necessary to do further research to investigate the truth. The similar disputes also exist in the research of the influences of microbubble density and initial velocity on drag reduction. In view of this, the influences of microbubble size, microbubble density, and microbubble initial velocity on drag reduction will be investigated again by the approach of model simulation in this paper. From the investigation results, it is found that the increasing microbubble density and initial velocity will enhance drag reduction. However, the microbubble size, varying from 0.01 mm to 0.2 mm, does not have significant influence on drag reduction. This finding coincides with the conclusion drawn in [26]. The details of the research are reported below.

2. MODELLING OF A SINGLE HULL VESSEL

In order to concentrate the research on investigating the effect of microbubbles on drag reduction, a single hull flat bottom vessel, rather than a SWATH type vessel, is considered in this paper. Moreover, the sea waves are also ignored for simplifying model simulation. In this section, the numerical model of a single hull vessel is established in ANSYS Fluent 18.1, see Figure 1. Its parameters are listed in Table 1.

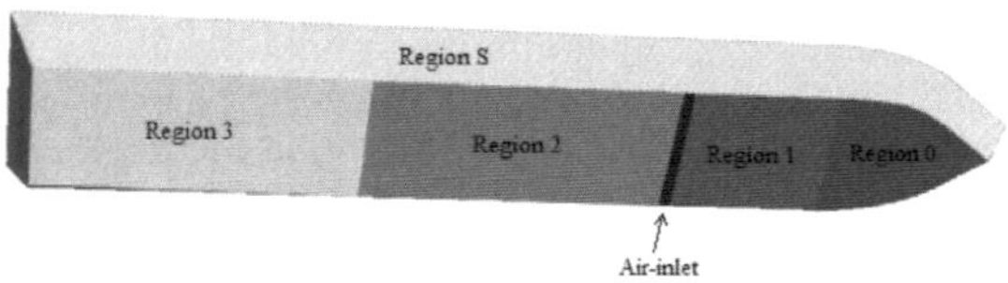

Figure 1: Model of the single hull flat bottom vessel

Table 1. Parameters of the vessel

Length (m)	**Width** (m)	**Draft** (m)	**Displacement** (kg)
1.80	0.26	0.12	39.80

In order to obtain a detailed understanding of the mechanism of microbubble technology, as shown in Fig.1 the vessel is divided into five sub-regions. They are Region S, Region 0, Region 1, Region 2, and Region 3, respectively. Among them, Region 0 is a 0.3 m long area starting from the bow of the vessel; Region 1 is also a 0.3 m long area following Region 0; Region 1 is followed by Region 2 and Region 3, both of which are 0.6 m long; Region S is the side of the vessel. The air-inlet locates between Region 1 and Region 2, and its size is 0.02 m × 0.26 m.

Assume the flow around the vessel is symmetric, attributing to which only half of the model is used to establish the domain for simulation calculations. Consequently, the number of meshes in model discretization as well as the burden of numerical calculation can be cut half. The domain that is defined for implementing the calculation is a rectangular parallelepiped area. Its entrance of watershed is at the upstream of the bow; its exit is at the downstream of the stern of the vessel. Its detailed parameters are listed in Table 2.

Table 2. Parameters of the domain for simulation calculations

Distance ahead of the vessel (m)	**Distance behind the vessel** (m)	**Distance below the vessel** (m)	**Distance above the vessel** (m)
1.80	1.80	1.80	5.40

Since mesh quality is critical for guaranteeing the accuracy of numerical calculation results, software ANSYS Fluent provides three important criteria for assessing the quality of meshing. They are skewness, aspect radio, and cell size change. For three-dimensional meshing, the skewness of the hexahedron should be less than 0.8 and the skewness of the tetrahedron should be less than 0.9. For the aspect ratio, different mesh elements will be assessed by using different calculation methods. The unit with an aspect ratio equal to 1 is regarded as the best unit, such as an equilateral triangle, a regular quadrilateral, a regular tetrahedron, a regular hexahedron, etc. In general, the aspect ratio should not be more than 5:1. The ratio of adjacent cell sizes is only applicable to 3D cells, which is preferably between 1 and 2. Following these rules, the model of the vessel is meshed. The meshing result of the entire calculation domain is shown in Figure 2. Where, the total number of grids is 26,000; the corresponding skewness of the mesh is less than 0.9; the aspect ratio of the mesh in the paper is less than 5:1; the ratio of adjacent cell sizes in the paper is less than 2. The most important is that there is no grid cell has negative volume and negative area, so the meshing results can fully meet the requirement of calculation accuracy.

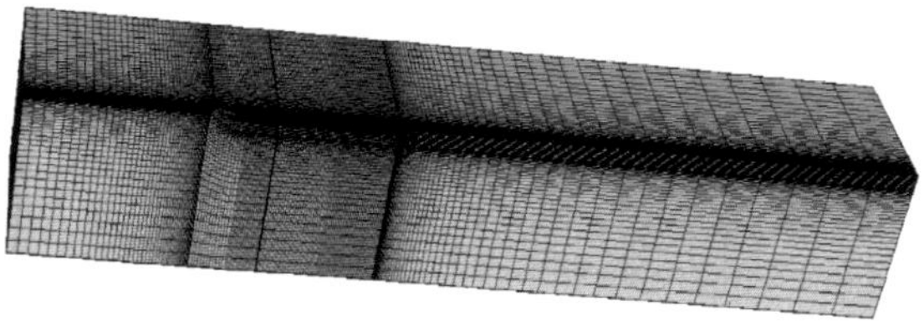

Figure 2: Meshing result of the entire calculation domain

In the following calculations, the vessel speed is set to be 1 m/s and the corresponding Reynolds number is equal to 1.69×10^6. When using the wall function method to deal with near-wall problems, the distance between the first layer of mesh and the wall is very important. In ANSYS Fluent, it is indicated by value Y^+, i.e.

$$Y^+ = \frac{\rho \mu_\tau y_b}{\mu} \tag{1}$$

According to (1), the distance between the first layer of mesh and the surface of vessel model is calculated. Its value is in the range of 100 to 180. Before conducting simulation calculation, the boundary conditions are defined in Figure 3. They include inlet, air inlet, pressure outlet, and four walls.

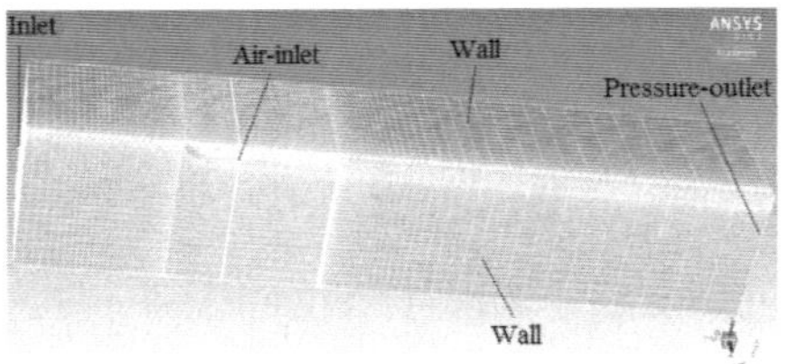

Figure 3: Boundary conditions of the calculation domain

More details of the boundary conditions are described as follows:

- The surface of the model is set as solid wall;
- Considering the symmetry of the model, the mid-longitudinal section of the model is taken as the symmetry plane boundary condition. At this point, the velocity along the Z axis is zero, i.e. P = 0;
- The inlet of the water flow is taken as the velocity inlet, the velocity direction is perpendicular to the boundary surface, i.e. $U_{inlet} = U_{\infty}$, $\gamma = P = 0$. Herein, U_{∞} is the velocity of main flow. The downstream exit of the stern is set as the boundary condition of the pressure outlet;
- The side and bottom of the calculation domain are far away from the model of the vessel, thus satisfying the condition of no-slip wall boundary;
- The side and bottom boundaries of the calculation domain are also taken as velocity inlet boundaries, i.e. $U_{inlet} = U_{\infty}$, $\gamma = P = 0$;
- The outlet of the calculation domain is taken as the pressure outlet. The static pressure of water is zero, and the pressure of the entire outlet boundary can be defined as the standard atmospheric pressure;
- When there is no bubble is generated and applied, the air inlet is set as a solid wall and the no-slip condition is satisfied.

3. INFLUENCE OF MICROBUBBLE SIZE

In order to quantitatively assess the influence of microbubbles on drag reduction, a criterion, namely drag reduction ratio r, is defined in this paper, i.e.

$$r = \frac{D_0 - D_1}{D_0} \times 100\% \quad (2)$$

Assume the vessel speed is 1 m/s, the density parameter ϵ of microbubbles remains 1, and the size of microbubbles is respectively 0, 0.01, 0.05, 0.07, 0.1, and 0.2 mm. Size 0 mm means that there is no microbubble is applied. The hull surface microbubble volume fraction distributions in different scenarios are calculated and the results are shown in Figure 4.

From Figure 4, it is difficult to see any obvious difference of different scenarios in microbubble volume fraction distributions. This seems indicate the insignificant influence of microbubble size on drag reduction. To confirm this finding from observation, the drag forces in different regions of the vessel in different scenarios were calculated and all calculation results are listed in Table 3.

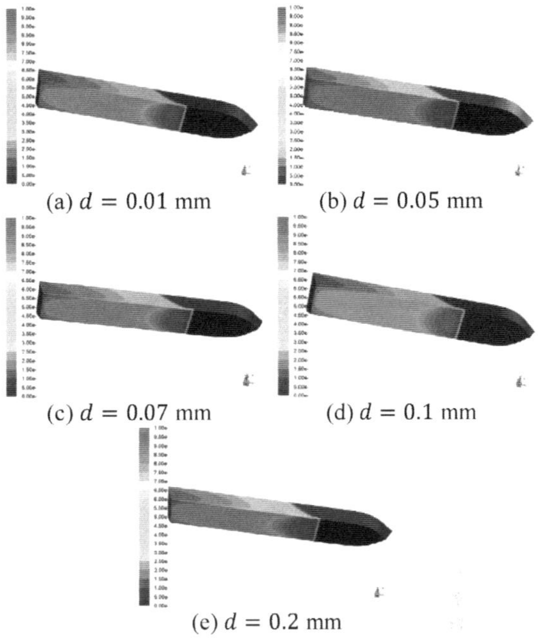

Figure 4: Hull surface microbubble volume fraction in different microbubble size scenarios

Table 3. Drag forces obtained in different microbubble size scenarios (N)

Region	Diameter of microbubble d (mm)					
	0	0.01	0.05	0.07	0.1	0.2
S	0.336	0.349	0.349	0.348	0.348	0.347
0	0.058	0.057	0.057	0.057	0.057	0.057
1	0.045	0.038	0.038	0.038	0.038	0.038
2	0.137	0.050	0.050	0.049	0.049	0.046
3	0.143	0.117	0.116	0.116	0.115	0.112
Total	0.719	0.611	0.610	0.609	0.607	0.600

From Table 3, it is interestingly found that despite the size of the microbubbles, they have different effects on the drag reduction in different sub-regions of the vessel. In other words, after applying microbubbles to the bottom of the vessel they increase the drag force in Region S while reduce the drag forces in the other sub-regions to varying degrees. Moreover, the calculation results of the total drag force indicate that despite the different effects of the microbubbles in different sub-regions, the application of them is generally helpful to reduce the drag of the vessel. In order to further demonstrate this finding intuitively, the drag reduction ratios r in different sub-regions in different microbubble size scenarios are calculated using (2). The corresponding results are listed in Table 4.

From Table 4, it is seen that compared with the other sub-regions, the drag force in Region 2 is most reduced by the microbubbles, then followed by Region 3 and Region 1. The drag force in Region 0 is also reduced by the microbubbles but the reduction is not very significant. The overall contribution of the microbubbles to drag reduction is over 15%. However, in terms of the influence of microbubble size the further observation of Tables 3 and 4 discloses that:

Table 4. Drag reduction ratio r based on the results in Table 3

Region	Drag reduction ratio r (%)					
	d=0 mm	d=0.01 mm	d=0.05 mm	d=0.07 mm	d=0.1 mm	d=0.2 mm
S	0	-3.71	-3.65	-3.60	-3.51	-3.22
0	0	1.25	1.25	1.24	1.24	1.22
1	0	14.24	14.21	14.19	14.14	13.98
2	0	63.53	63.84	64.12	64.66	66.72
3	0	18.63	19.02	19.34	19.93	21.69
Total	0	15.10	15.26	15.40	15.66	16.53

- In the regions ahead of the air inlet (i.e. Region 0 and Region 1), the larger the size of microbubbles, the less the drag force is reduced;
- In the regions behind the air inlet (i.e. Region 2 and Region 3), the larger the size of microbubbles, the more the drag force is reduced;
- In terms of the overall drag reduction, the larger the size of microbubbles, the more the drag force is reduced;
- But despite the above tendencies, in the range of diameter 0~0.2 mm the change of microbubble size has insignificant influence on the drag reduction in all 5 regions.

4. INFLUENCE OF MICROBUBBLE DENSITY

The influence of microbubble density on drag reduction will be investigated in this section. In the investigation, it is assumed that the microbubble diameter is 1 mm and the microbubble initial velocity is 1 m/s. The density parameter ϵ of the microbubbles is set to be 0.01, 0.1, 0.5, 0.7, and 1, respectively. Likewise, the hull surface microbubble volume fraction distributions in different microbubble density scenarios are investigated first. They are illustrated in Figure 5.

From Figure 5, it is clearly seen that the variation of microbubble density significantly changes the microbubble volume fraction distribution on the vessel surface. This implies that microbubble density may significantly influence the drag reduction. Hence, the drag forces and the corresponding drag reduction ratio r in different microbubble density scenarios are calculated. The results are listed in Table 5. Where, $\epsilon = 0$ means that there is no microbubble is applied to the bottom of the vessel.

From Table 5, it is found that with the increase of microbubble density, the drag force either increases or decreases slightly in Region S. In Region 0, the application of microbubbles increases drag reduction, however the change of microbubble density does not significantly affect drag reduction. But in the other 3 regions (i.e. Region 1, Region 2 and Region 3), the increase of microbubble density does significantly improve drag reduction in particular in Region 2. In terms of the total drag force, with the increase of microbubble density the drag reduction increases as well. When the microbubble density parameter $\epsilon = 1$, the drag reduction ratio r can reach a value as large as 15.10%. From these calculation results, it can be concluded that the change of microbubble density does have a significant influence on drag reduction. Moreover, in the range of $\epsilon = 0 \sim 1$ the higher the microbubble density, the more the drag will be reduced.

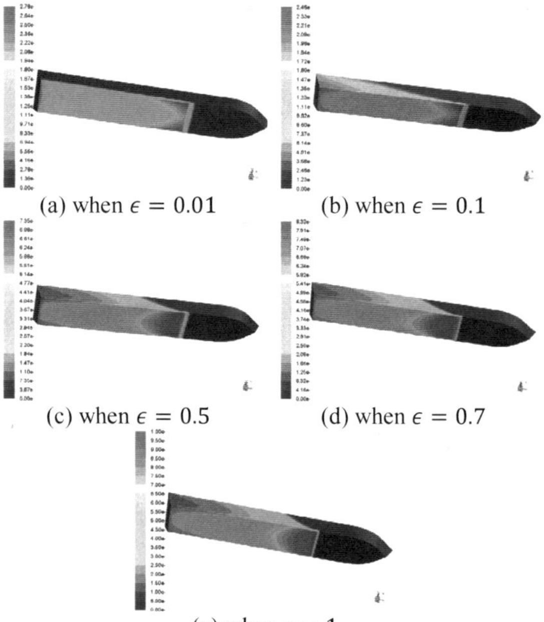

(a) when $\epsilon = 0.01$ (b) when $\epsilon = 0.1$ (c) when $\epsilon = 0.5$ (d) when $\epsilon = 0.7$ (e) when $\epsilon = 1$

Figure 5: Hull surface microbubble volume fraction in different microbubble density scenarios

Table 5. Drag forces and the corresponding r in different microbubble density scenarios

Region	Drag force (N)	r (%)	Drag force (N)	r (%)	Drag force (N)	r (%)
	$\epsilon = 0$		$\epsilon = 0.01$		$\epsilon = 0.1$	
S	0.336	0	0.334	0.63	0.347	-3.25
0	0.058	0	0.058	0.09	0.058	0.07
1	0.045	0	0.044	0.38	0.044	1.84
2	0.137	0	0.137	0.16	0.118	13.89
3	0.143	0	0.141	1.75	0.137	4.17
Total	0.719	0	0.714	0.71	0.704	2.08
	$\epsilon = 0.5$		$\epsilon = 0.7$		$\epsilon = 1$	
S	0.357	-6.30	0.354	-5.40	0.349	-3.71
0	0.057	0.71	0.057	0.94	0.057	1.25
1	0.041	8.47	0.040	11.05	0.038	14.24
2	0.069	49.74	0.059	57.33	0.050	63.53
3	0.120	16.58	0.118	18.00	0.117	18.63
Total	0.644	10.44	0.628	12.78	0.611	15.10

5. INFLUENCE OF MICROBUBBLE INITIAL VELOCITY

Finally, the influence of the initial velocity of microbubbles on drag reduction is investigated in this section. In the investigation, it is assumed that the microbubble diameter is 1 mm, the microbubble density parameter $\epsilon = 1$, the initial velocity of the microbubbles v is set to be 0.1, 0.5, 1, 5, 10 m/s, respectively. The time step is 0.01 second, and the step number is 1000. The hull surface microbubble volume fraction distributions obtained in different initial velocity scenarios are shown in Figure 6.

(a) when $v = 0.1$ m/s

(b) when $v = 0.5$ m/s

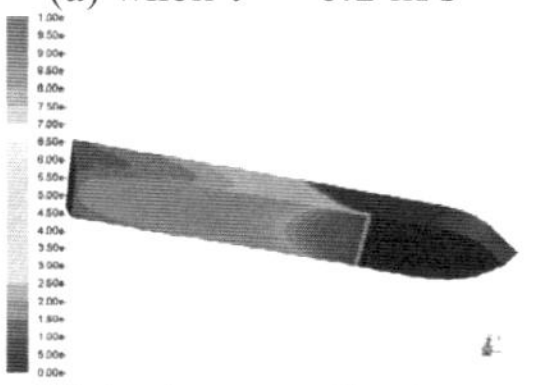

(c) when $v = 1$ m/s

(d) when $v = 5$ m/s

(e) when $v = 10$ m/s

Figure 6: Hull surface microbubble volume fraction in different initial velocity scenarios

From Figure 6, it is clearly seen that the initial velocity of microbubbles can significantly change the microbubble volume fraction distribution on the hull surface of the vessel. So, it can be predicted that it will also affect a lot drag reduction. Similar to those observed from Figs. 4 and 5, the largest volume fraction of microbubble is present at the position of air inlet, then decreases gradually along the flow direction. The further observation of Fig.6 discloses that with the increase of the initial velocity, stable microbubble layer is established in a larger area. From this phenomenon, it can be inferred that in the range of the initial velocities being considered the higher the initial velocity, the more the drag force will be reduced. In order to confirm this prediction, the drag forces and the corresponding drag reduction ration r in different initial velocity scenarios are calculated. The results are listed in Table 6. Where, initial velocity $v = 0$ m/s means that there is no microbubble is applied to the vessel.

From Table 6, it is found that when the initial velocity of the microbubbles is low, the microbubbles in Region S increase the friction between the hull and water (thus drag), so the drag reduction is negative. But with the continual increase of the microbubble initial velocity, the microbubbles will change their role in drag reduction, i.e. start to increase drag reduction in Region S. This is because stable bubble layer can be established on the side surface of the vessel only when the microbubble initial velocity is large enough, see Figures 6d and e. By contrast, the drag reduction in Region 0 is not significantly affected by the change of microbubble initial velocity because few microbubbles can reach Region 0 due to the flow direction and its far distance to air inlet. But in the other 3 regions (i.e. Regions 1-3) particularly in Region 2, the drag reduction rate r rapidly increases with the increase of microbubble initial velocity. When the initial velocity is 5 m/s, the drag reduction ratio r in Region 2 has reached 72.19%. However, when the initial velocity is 10 m/s, the drag reduction ration r in both Region 2 and Region 3 decreases a little bit, i.e. from 72.19% drops down to 68.87% in Region 2 and from 20.19% down to 15.33% in Region 3, respectively. This is because when the microbubble velocity is very high (e.g. 10 m/s), some microbubbles will be blown directly to the stern of the vessel. As a consequence, no all microbubbles will contribute to the construction of stable air layer in this case, thus leading to smaller value of drag reduction ratio r. Such an explanation can be proved by Figure 7, where the microbubble volume fraction distributions in Region 2 and Region 3 when the microbubble initial velocity is respectively equal to 5 m/s and 10 m/s are illustrated.

Table 6. Drag forces and corresponding r obtained in different initial velocity scenarios

Region	Drag force (N)	r (%)	Drag force (N)	r (%)	Drag force (N)	r (%)
	$v = 0$ m/s		$v = 0.1$ m/s		$v = 0.5$ m/s	
S	0.336	0	0.346	-2.84	0.358	-6.35
0	0.058	0	0.058	0.18	0.057	0.70
1	0.045	0	0.044	1.91	0.041	8.41
2	0.137	0	0.119	13.26	0.069	49.53
3	0.143	0	0.137	4.84	0.120	16.53
Total	0.719	0	0.703	2.30	0.645	10.37
	$v = 1$ m/s		$v = 5$ m/s		$v = 10$ m/s	
S	0.349	-3.71	0.328	2.32	0.295	12.24
0	0.057	1.25	0.056	3.21	0.054	5.63
1	0.038	14.24	0.033	25.58	0.027	39.45
2	0.050	63.53	0.038	72.19	0.043	68.87
3	0.117	18.63	0.114	20.19	0.121	15.33
Total	0.611	15.10	0.570	20.74	0.541	24.83

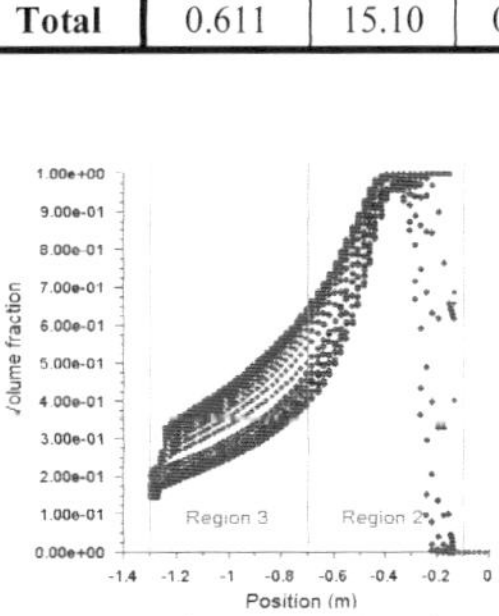

(a) when $v = 5$ m/s

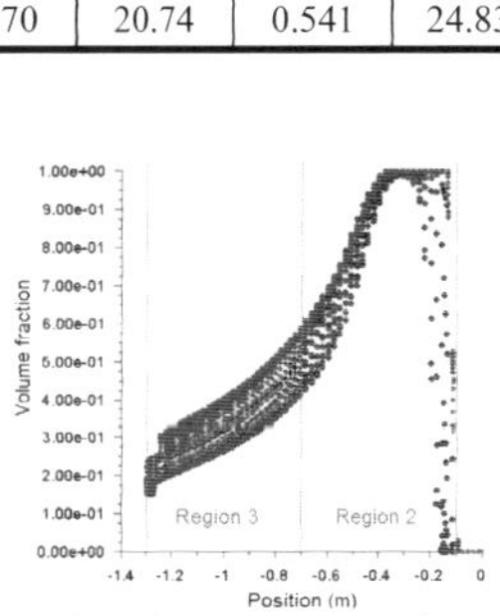

(b) when $v = 10$ m/s

Figure 7: Volume fraction distribution of microbubbles in different initial velocity scenarios

From Figure 7, it is clearly seen that more microbubbles are present in both Region 2 and Region 3 when the initial velocity $v = 5$ m/s than in the case when $v = 10$ m/s. It has no doubt that the more the microbubbles are present, the easier the stable air layer can be formed in the regions. This explains why the values of the drag reduction ratio r in both Regions 2 and 3 decrease to certain extent when microbubble initial velocity increases from 5 m/s to 10 m/s.

6. CONCLUSIONS

In order to reach a consensus on microbubbles' effect on drag reduction, the influences of microbubble size, microbubble density, and microbubble initial velocity on drag reduction are investigated in this paper by the approach of numerical simulation. From the calculation results, the following conclusions can be drawn:

- Drag reduction is highly dependent on the volume fraction distributions of microbubbles on the surface of the vessel. In principle, the more microbubbles are present, the more drag force will be reduced. Since the microbubble volume fraction is not equally distributed on the surface of the vessel, the obtained drag reduction ratios are different in different sub-regions of the vessel. For example, since the largest microbubble volume fraction always appears in Region 2, so the drag force in Region 2 is most reduced by microbubbles;
- The application of microbubbles does reduce drag force of the vessel. However, the drag reduction is not very sensitive to the change of microbubble size;
- The change of microbubble density can significantly influence the drag reduction. Moreover, the higher the microbubble density, the more the drag force will be reduced;
- Drag reduction is also sensitive to the initial velocity of microbubbles. In other words, the increasing microbubble initial velocity will increase drag reduction. However, the amount of drag reduction would be capped or even decrease when the initial velocity is very high (e.g. 10 m/s). This is because part of microbubbles is blown to the stern of the vessel in this case. As a consequence, these microbubbles will not contribute to the construction of air layer.

7. REFERENCES

1. Molland, A. F., Turnock, S. R. and Hudson, D. A., 'Ship resistance and propulsion', *Cambridge University Press*, 2017.
2. Phillips, S., Shin, I. B. and Armstrong, C., 'Crew transfer vessel performance evaluation', *Proceedings of Design & Operation of Wind Farm Support Vessels Conference*, 28-29 January 2015, London, UK.
3. What is a SWATH ship? Available at https://www.marineinsight.com/types-of-ships/what-is-a-swath-ship. Recently accessed on January 10, 2019.
4. https://www.thecrownestate.co.uk/en-gb/media-and-insights/news/2018-the-crown-estate-shares-further-detail-on-plans-for-round-4-including-proposed-locations-to-be-offered-for-new-seabed-rights. Recently accessed on January 10, 2019.
5. Baba, E., 'A new component of viscous resistance of ships', *Journal of the Society of Naval Architects of Japan*, 125, 1969, pp. 23-34.
6. Fukuda, K., Tokunaga, J., Nobunaga, T., Nakatani, T. and Iwasaki, T., 'Frictional drag reduction with air lubricant over a super-water-repellent surface', *Journal of Marine Science and technology*, 5(3), 2000, pp. 123-130.
7. Ceccio, S. L., Perlin, M. and Elbing, B. R., 'A cost-benefit analysis for air layer drag reduction', *International Conference on Ship Drag Reduction SMOOTH-SHIPS*, Istanbul, Turkey, 20-21 May 2010, Paper No.4.
8. Jang, J., Choi, S. H., Ahn, S. M., Kim, B. and Seo, J. S., 'Experimental investigation of frictional resistance reduction with air layer on the hull bottom of a ship', *International Journal of Naval Architecture and Ocean Engineering*, 6, 2014, pp. 363-379.
9. Mohammad, A. and Majid, M., 'A review on the drag reduction methods of the ship hulls for improving the hydrodynamic performance', *International Journal of Maritime Technology*, 4, 2015, pp. 51-64.
10. Hashim, A., Yaakob, O. B., Koh, K. K., Ismail, N. and Ahmed, Y. M., 'Review of micro-bubble ship resistance reduction methods and the mechanisms that affect the skin friction on drag reduction from 1999 to 2015', *Journal Teknologi (Science & Engineering)*, 74(5), 2015, pp. 105-114.
11. Ceccio, S. L., 'Friction drag reduction of external flows with bubble and gas injection', *Annual Review of Fluid Mechanics*, 42, 2010, pp. 183–203.
12. Jacob, B., Olivieri, A., Miozzi, M., Campana, E. F. and Piva, R., 'Drag reduction by microbubbles in a turbulent boundary layer', *Physics of Fluids*, 22, 2010, pp. 115104.
13. Shen, X., Ceccio, S. L. and Perlin, M., 'Influence of bubble size on microbubble drag reduction', *Experiments in Fluids*, 41(3), 2006, pp. 415-424.
14. Murai, Y., Oishi,Y. and Yamamoto, F., 'Turbulent shear stress profiles in a bubbly channel flow assessed by particle tracking velocimetry', *Experiments in Fluids*, 41(2), 2006, pp. 343–352.
15. Van den Berg, T. H., Luther, S., Lathrop, D. P. and Lohse, D., 'Drag reduction in bubbly Taylor-Couette turbulence', *Physical Review Letters*, 94, 2005, pp. 044501.
16. Gabillet, C., Collin, C. and Fabre, J., 'Experimental study of bubble injection in a

turbulent boundary layer', *International Journal of Multiphase Flow*, 28(4), 2002, pp. 553-578.
17. Kodama, Y., Kakugawa, A., Takahashi, T. and Kawashima, H., 'Experimental study on microbubbles and their applicability to ships for skin friction reduction', *International Journal of Heat and Fluid Flow*, 21, 2000, pp. 582-588.
18. Latorre, R., 'Ship hull drag reduction using bottom air injection', *Ocean Engineering*, 24(2), 1997, pp. 161-175.
19. Ortiz-Villafuerte, J. and Hassan, Y. A., 'Investigation of microbubble boundary layer using particle image velocimetry', *ASME Journal of Fluids Engineering*, 128(3), 2006, pp. 507–519.
20. Lu, J., Fernandez, A. and Tryggvason, G., 'The effect of bubbles on the wall drag in a turbulent channel flow', *Physics of Fluids*, 17, 2005, pp. 95–102.
21. McCormick, M. E. and Bhattacharyya, R., 'Drag reduction of a submersible hull by electrolysis', *Naval Engineers Journal*, 85(2), 1973, pp. 11–16.
22. Kawamura, T. and Kodama, Y., 'Numerical simulation method to resolve interactions between bubbles and turbulence', *International Journal of Heat and Fluid Flow*, 23(5), 2002, pp. 627–638.
23. Kawamura, T., Fujiwara, A., Takahashi, T., Kato, H., Matsumoto, Y. and Kodama, Y, 'The effects of the bubble size on the bubble dispersion and skin friction reduction'. National Maritime Research Institute, Japan. Available at http://www.nmri.go.jp/oldpages/turbulence/PDF/symposium/FY2003/Kawamura.pdf. Recently accessed on January 10, 2019.
24. Jones, S. F., Evans, G. M. and Galvin, K. P., 'Bubble nucleation from gas cavities – A review', *Advances in Colloid and Interface Science*, 80(1), 1999, pp. 27–50.
25. Xu, J., Maxey, M. R. and Karniadakis, G. E., 'Numerical simulation of turbulent drag reduction using micro-bubbles', *Journal of Fluid Mechanics*, 468, 2002, pp. 271-281.
26. Kawamura, T., Moriguchi,Y., Kato, H., Kakugawa, A. and Kodama, Y., 'Effect of bubble size on the microbubble drag reduction of a turbulent boundary layer', *Proceedings of ASME/JSME The 4th Joint Fluids Summer Engineering Conference (FEDSM2003)*, Honolulu, Hawaii, USA, July 6-10, 2003.

8. AUTHORS BIOGRAPHY

Wenxian Yang holds the current position of Senior Lecturer in Offshore Renewable Energy at the School of Engineering of Newcastle University. He is responsible for developing and teaching the advanced offshore renewable energy and the relevant operation and maintenance techniques, including the development of the new concept of offshore wind farm service vessels. His previous experience includes wind turbine condition monitoring and fault diagnosis, SCADA data analysis, scour protection of offshore structures, the application of biomimetic techniques in the design of wind turbine, etc.

Hongwei Yang is a postgraduate student who is studying in the School of Engineering of Newcastle University. In this paper, he is responsible to develop the numerical model of the single hull vessel and conduct simulation calculations in ANSYS Fluent.

Ji Wang is a Research Engineer in Mingyang Smart Energy Group Ltd based in Shanghai China. In this paper, he works together with Mr Hongwei Yang to develop the numerical model of the single hull vessel and conduct simulation calculations in ANSYS Fluent.

Meiping Sheng holds the current position of Professor at the School of Marine Science and Technology of Northwestern Polytechnical University (NPU). She chairs the offshore renewable energy research at NPU. In addition to developing innovative offshore renewable technologies, her research interest also includes structure vibration control, testing, structural strength analysis, and optimization design. Professor Sheng is a winner of the China Youth Science and Technology award, the award of Chinese young women scientists and many other provincial and ministerial level awards in science and technology. In this paper, she provides technical help in simulating the microbubble technique in ANSYS Fluent.

Minqing Wang holds the current position of Professor at the School of Marine Science and Technology of Northwestern Polytechnical University. In this paper, he comments the simulation calculation results and also makes contribution during the period of developing this paper.

TARGETS NOT INCLUDED!

SY Heriot, D McKnight & B Furdek, Wärtsilä Guidance Marine, UK
H Busshoff, Bernhard Schulte Ship Management, Germany

SUMMARY

The latest technology in local position reference sensing is changing the way that vessels operate alongside assets, whether that be an oil platform, another vessel or an off-shore wind turbine. These technologies allow local positioning to be completely independent of the asset which enables safer, more efficient operations as well as reducing costs for the asset owner. SceneScan, launched in 2018, is the first laser local position reference sensor that does not require a physical reflective target to be placed on the asset to operate; this offers many operational advantages. It follows the success of the first targetless sensor, RangeGuard Monopole which is used for SOV operations at offshore wind farms, launched in 2016 and uses radar technology. This paper describes SceneScan Monopole which is a special feature included with the standard SceneScan software that allows SceneScan to be used for SOV operations at a windfarm. With the launch of SceneScan Monopole, Wärtsilä Guidance Marine can now offer a completely targetless solution for SOV's. Using RangeGuard Monopole alongside SceneScan Monopole provides two forms of redundancy (laser and radar), and is not reliant on GPS, fulfilling DP2 requirements.

NOMENCLATURE

GPS	Global positioning system
DGPS	Differential global positioning system
GNSS	Global navigation satellite system
DP	Dynamic positioning
PRS	Position reference sensor(s)
SOV	Service Operations Vessel

1. INTRODUCTION

Everyone is familiar with GNSS for navigation (GPS). Vessels that visit oil platforms use the more accurate DGPS for positioning close to them. However GPS is known to be unreliable. The Royal Academy of Engineering lists no less than 24 failure modes which in the most dramatic cases can result in 1000's of meters in error or total loss of signal [1]. The same can happen at the oil platform or wind turbine, so local positioning systems are needed to be used alongside DGPS to ensure that the manoeuvres are completely safe and precise and positioning is not reliant on just one type of sensor.

Traditional local relative positioning sensors such as CyScan AS [2] or RadaScan View [3] rely on placing targets on the structure which the sensor on the vessel can detect and report its position relative to. This position information is fed into the DP system of the vessel, enabling it to maintain position near the asset without having to rely on DGPS.

2016 saw the launch of RangeGuard Monopole, the first DP reference sensor that allowed an SOV to approach an offshore wind tower that did not need any targets installing [4,5]. This sensor was the first of its kind based on radar technology.

In 2018, SceneScan was launched [6]. This is the first targetless local PRS that is based on laser technology. SceneScan uses natural or man-made structures within the sensor field of view as its target and the latest software algorithms developed allows it to be used against monopole structures such as offshore wind turbines.

2. WHY TARGETLESS TECHNOLOGY?

It is common to hear reports from the crews of SOV's that traditional laser systems (designed for use at oil platforms or other large assets) that rely on a reflective target are not optimised for the wind farm environment [7].

- Reflective targets are poorly placed.
- Targets are obscured by the walk-to-work gangway.
- Targets are of poor quality, using low cost low grade reflective tape rather than high quality reflective prisms [8]. Low quality target reflectivity will degrade significantly over time leading to a reduction in operating range.
- Targets can be mistaken for hi-vis jackets of workmen on the walk-to-work gangway or on the wind tower platform. This may have serious consequences if the sensor locks onto a moving person rather than a fixed target.

Any of these issues could result in a catastrophic loss of position and should be taken extremely seriously. When these systems are in use is it highly likely that people will be walking on the gangway, and so physical separation of reflector from people is not possible by design (unlike on an oil platform).

As well as safety, other benefits for targetless sensors include

- No maintenance costs for targets.
- No expenditure on targets (each wind tower requires at least 2 targets with a traditional system).
- Faster vessel turnaround times.

- Vessel independent of asset.
- Positioning of the ship is faster. The repeatability of a position is better because turbine geometry is usually the same, whereas targets may be installed in slightly different orientations on different turbines.
- Less protocols and approval – no need ask for approval to install targets.
- Operational advantages – not having to request for funds for targets.

3. THE SCENESCAN MONOPOLE SYSTEM

3.1 INTRODUCTION TO SCENESCAN

SceneScan looks very much like the predecessor to CyScan AS, the CyScan MK4 [9]. It uses the same form factor and much of hardware is similar, however the optics, optoelectronics and software algorithms differ significantly.

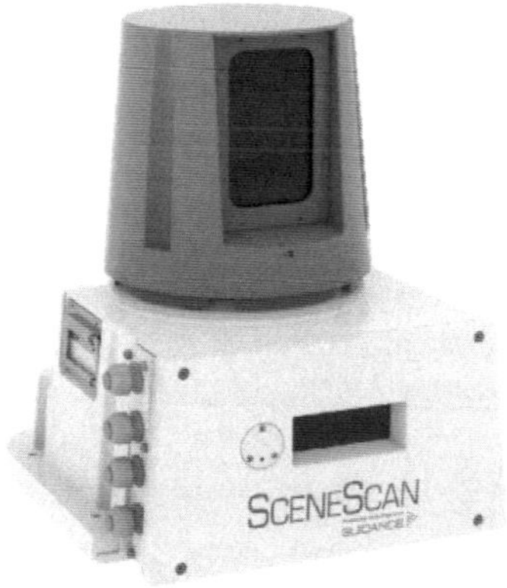

Figure 1. SceneScan

Whereas CyScan detects and tracks position from discrete reflective targets, such as a rugged prism [10], SceneScan provides tracking information relative to natural or man-made structures within the sensor field of view.

It does this by first creating a 3D scan of the scene; building it out of a series of 2D slices. This process typically takes a minute. It then finds the most suitable 2D slice to track off by selecting the one which appears to be the most reliable and informative among all the 2D sections captured in the previous step.

Once a suitable 2D slice has been selected, the system starts tracking and reporting the position of the sensor within the scene. As the vessel moves and the view of the scene changes, the sensor will automatically update its model of the scene while tracking; all without any need for user intervention.

By default, the reference point to where range, bearing and heading is measured is the centre of the 2D point cloud, but the user can also select the reference point manually if desired [11].

A full description of how the SceneScan tracks and its performance compared to other reference systems is described in 'SceneScan Tracking and Performance' [12] and 'How targetless local position reference sensors will change and improve the experience of DP Operators' [13].

Figures 2 and 3 show some examples of the SceneScan in action.

Figure 2 (a). SceneScan was used at a drillship in Aberdeen harbour.

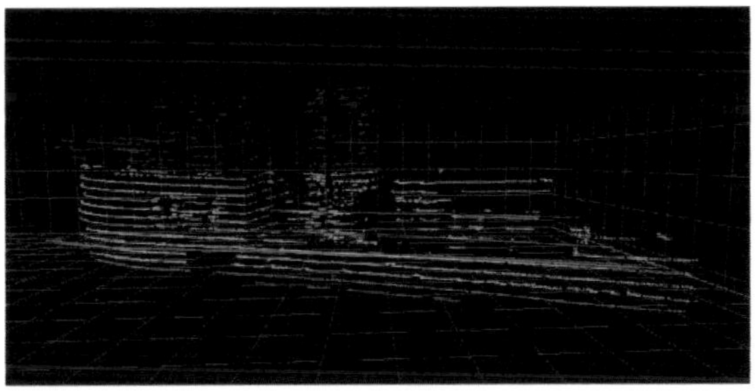

Figure 2 (b). 3D scan of drill ship. The scan is made up of slices which are colour coded. Green slices are suitable for tracking, red is not suitable, orange slices are borderline, and blue is ignored due to the high sensor tilt angle required.

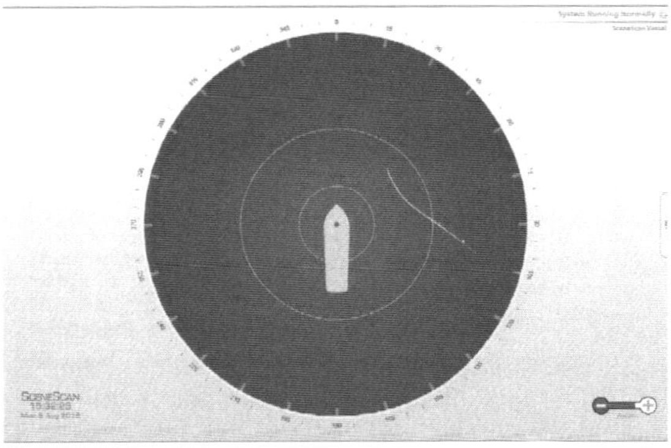

Figure 2 (c). User dashboard interface used by the DPO.

The sensor is represented by the red dot within the vessel schematic in the centre of the display. The 2D scan of the drill ship (at the selected tilt angle) is to the right of the sensor.

Figure 3 (a). SceneScan was used against some silos at Aberdeen Harbour.

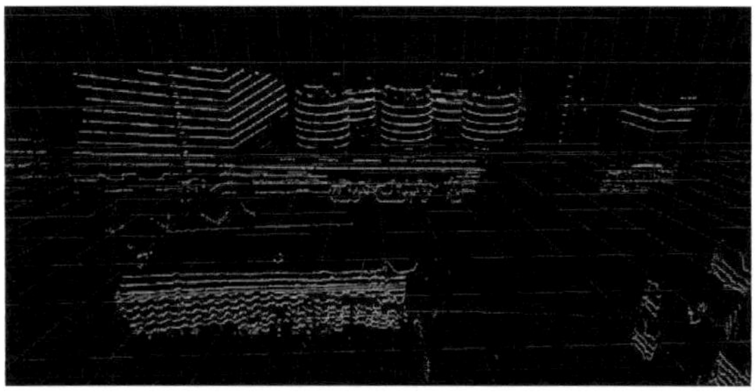

Figure 3 (b). 3D scan of Silos.

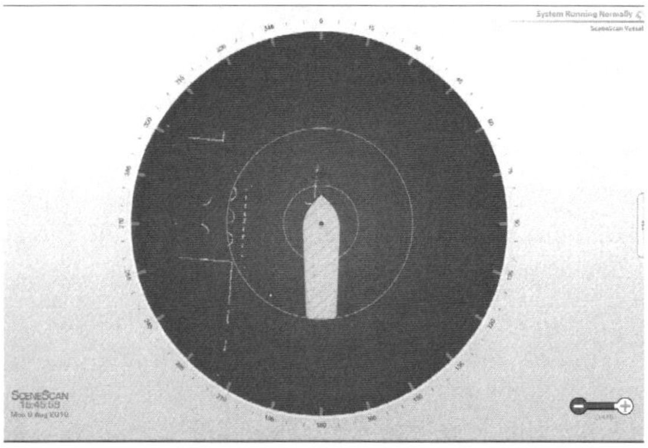

Figure 3 (c). User dashboard interface which is used by the DPO.

It is apparent from figures 2 and 3 that SceneScan needs to 'see' several discreet features in the surrounding environment to enable tracking.

SceneScan relies on being able to recognise small changes in the scene as the vessel moves. Featureless structures such as a flat wall, or the side of an FPSO prove challenging for SceneScan. In these types of environments, if a vessel moves parallel to a very flat structure, the scene that it sees will not change. In this situation, the SceneScan will not find a suitable 2D slice of the scene and will not track.

A similar situation occurs when the structure in the environment is a single cylindrical object, such as a wind farm monopole. Although the SceneScan can see how far the structure is, it is possible to navigate around a monopole at a set distance and notice no difference in the scene.

3.2 SCENESCAN MONOPOLE

To enable targetless tracking at a single monopole, a new algorithm has been developed, known as Monopole mode. The Monopole mode is a standard feature of SceneScan and can be enabled by the user as required.

While the standard SceneScan tracking algorithm attempts to resolve the position and heading of the sensor, the Monopole algorithm calculates the position (centre point) and radius of the monopole.

Instead of using all points returned from the scene we instead specifically look for a single monopole feature which has a radius that lies within expected limits. Once we have detected a suitable monopole we can then track it. Any other data in the scene is ignored.

In the Monopole mode the Scene reference point is always at the centre of the monopole. This is shown on the user interface as a yellow dot and the system reports the range, bearing to the centre of the monopole and its radius. The edge of the monopole can also be seen in the user interface.

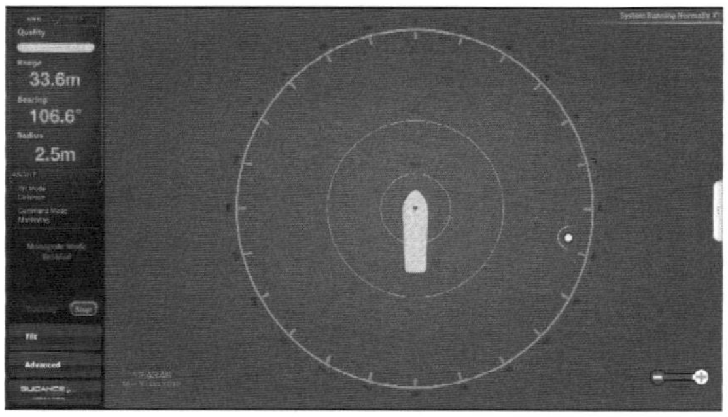

Figure 4. User dashboard interface in Monopole mode.

The sensor is represented by the red dot in the centre of the display. The yellow dot represents the centre of the monopole and the white arc of dots above it are the raw laser measurements showing the wall of the monopole.

The Monopole mode algorithm uses the widely known Hough transform method to find the centre point and the radius of the circle in the Scene. This is a suitable approach as a circle is the 2D representation of a single slice of the monopole structure.

4. SEA TRIALS

In August 2018 SceneScan was installed on the Bernhard Schulte managed Windea La Cour which operates at the Gemini offshore wind farm in the North Sea.

Figure 5. SceneScan on the Windea La Cour

For the first two weeks, the SceneScan was configured to operate in a simple data collection mode. During this time, there was no interaction from the crew of the vessel and the sensor automatically collected scanning data.

After this initial data collection period, the data was retrieved and replayed through the new Monopole algorithms. Amendments were made to the algorithms to improve accuracy and robustness of the reported position information.

In September a Wärtsilä Guidance Marine service engineer sailed with the Windea La Cour and the latest software was installed. The sensor was not connected to the DP system, however in all other respects it was fully operational and could be used in conjunction with the other DP reference sensors to report range and bearing. During this time data was collected for final analysis.

5. RESULTS

Figure 6 shows a scan of the monopole at close range during station keeping. The scale is in meters. It can be seen that the radius of the monopole is about 2.5m.

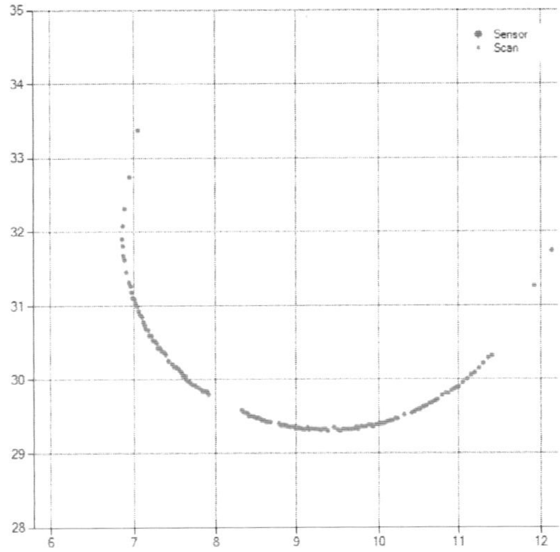

Figure 6. Close range scan of a monopole. Scale is in meters.

Figure 7 shows the approach of the vessel to a monopole. The monopole is detected at round 175m. The vessel approaches the turbine tracking all the way and holds station at around 30m. After a period, the vessel moves away. It holds reliably until a first loss of tracking at 235m. It continues to track with some disruptions out to almost 300m. This demonstrates extremely reliable performance.

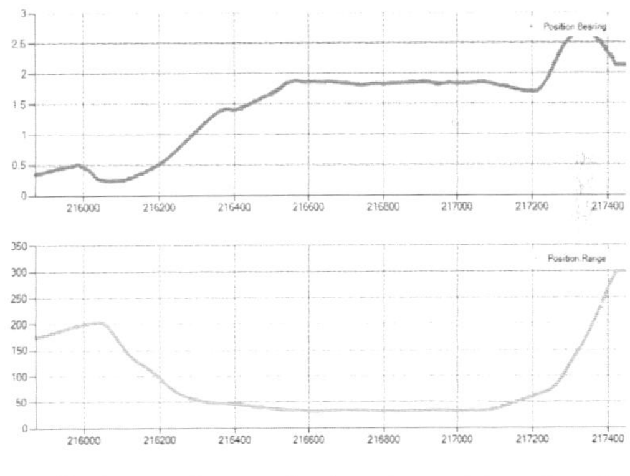

Figure 7. Approach to, station keeping and departure from a wind turbine. Top (blue) bearing in radians, bottom (yellow) range in meters.

Figure 8 is a zoomed in section of figure 7 during the station keeping period when the vessel is closest to the monopole. This shows that the noise generated by the SceneScan is very minimal

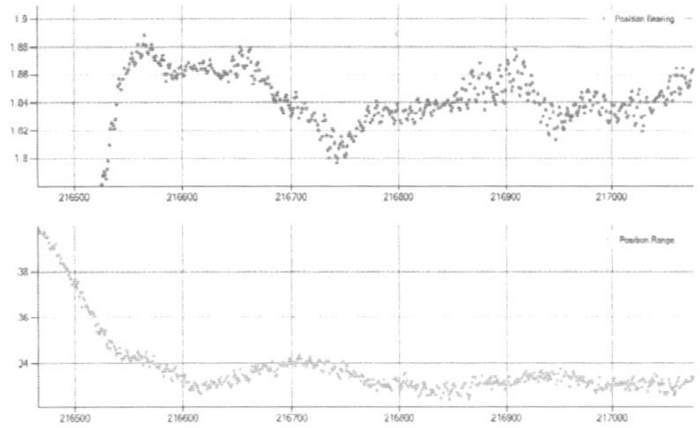

Figure 8. Zoomed in section of figure 7. Top (blue) bearing in radians, bottom (yellow) range in meters.

It is useful to look at even shorter timescales as these tracking graphs show a combination of the noise of the sensor and any real movement of the system.

Figure 9 shows a zoomed in section of figure 8. We can clearly see the wave motion of the vessel bobbing around. This observation of wave motion is excellent to see and is exactly what we would expect to see with all other traditional local position reference sensors.

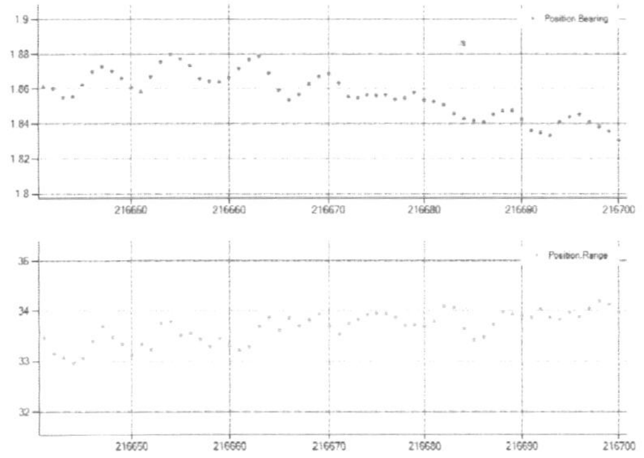

Figure 9. Zoomed in section of figure 8 shows the wave motion of the vessel. Top (blue) bearing in radians, bottom (yellow) range in meters.

It should be noted that range resolution for SceneScan is 8.5mm, noise resolution is 30mm and position noise when station keeping at TLP is less than 100mm.

6. DISCUSSION

When SceneScan Monopole was first presented to the crew of the Windea La Cour, it was initially met with some scepticism. The concept of a laser system with no targets is very radical, even though the crew are already familiar with targetless RangeGuard Monopole which they have been using for almost 2 years.

However, after explaining the differences with CyScan AS and SceneScan, any concerns quickly dissipated, and they were happy to trial it.

After a couple days of use, it was apparent that the crew really started to enjoy using SceneScan. It managed to track the monopole targets even in rough sea conditions.

On one day during the sea trials, the wave height was too large to deploy most of the workers at the turbines. Even in these harsh weather conditions the SceneScan was able to continue tracking. It was also observed by the crew that the SceneScan had very comparable range to the CyScan at the turbines.

The SceneScan will remain on the vessel for further testing and data collection and it is intended that the SceneScan will be on the vessel connected to the DP system when it shortly undergoes its annual FMEA testing.

7. CONCLUSIONS

SceneScan has already been proven as a targetless PRS for use against assets such as oil platforms. With the latest trials, it has now been demonstrated successfully for use against single monopole structures. The new monopole algorithm and features will be supplied as standard with SceneScan.

Used together, RangeGuard Monopole and SceneScan Monopole offer the first completely targetless solution for SOV's operating at offshore wind farms. With two forms of redundancy (radar and laser), this is the safest combination of local position reference sensors for vessel owners. This pairing has no need for targets, resulting in a significant cost saving for the wind farm operator / owner.

8. ACKNOWLEDGEMENTS

Aberdeen harbour board for giving permission to test and demonstrate SceneScan in the harbour

Crew of the Windea La Cour for cooperation during testing of the SceneScan

Ben Botkin, Wärtsilä Guidance Marine for assisting with sea trials and sailing with the Windea La Cour

9. REFERENCES

1. The Royal Academy of Engineering, Global Navigation Space Systems: reliance and vulnerabilities, *http://www.raeng.org.uk/publications/reports/global-navigation-space-systems*, 2011

2. CySca AS: http://www.guidance.eu.com/CyScan-AS

3. RadaScan View http://www.guidance.eu.com/radascan-view

4. SY Heriot, T Coggins, M Mundrinic, A Stead, H Busshoff, OM Husøy 'A New Era in Position Referencing', *Design & Construction of Wind Farm Support Vessels, London, UK*, 30-31 March 2016. http://www.guidance.eu.com/assets/_managed/cms/files/ANEWERAINLOCALPOSITIONREFERENCING_FINAL140316.pdf

5. A Knight, SY Heriot, H Busshoff, 'DP Positioning – Relative Positioning goes targetless', *Design & Construction of Wind Farm Support Vessels, London, UK*, 29-30 March 2017. http://www.guidance.eu.com/assets/_managed/cms/files/DPPositioning-RelativePositionReferencingGoesTargetless_final.pdf

6. SceneScan: http://www.guidance.eu.com/SceneScan

7. A Stead, H Busshoff, CR Grams, 'A Safety Related Approach to Wind Tower SOV Approach Instrumentation', *Design & Construction of Wind Farm Support Vessels, London, UK*, 29-30 March 2017. http://www.guidance.eu.com/assets/_managed/cms/files/HOMEIMPROVEMENTSTICKYBACKPLASTICFORWINDFARMSASAFETYRELATEDAPPROACHTOWINDTOWERSOVAPPROACHINSTRUMENTATION.pdf

8. Guidance Marine, 'PRS Reflector Target Recommendation for the Offshore Wind Industry', *Guidance Marine document 94-0499-4*. http://www.guidance.eu.com/assets/_managed/cms/files/PRSReflectorTargetRecommendationfortheOffshoreWindIndustry.pdf

9. CyScan: http://www.guidance.eu.com/cyscan

10. Rugged Prism https://marine.direct/index.php/platform-dp-targets/laser-targets/prism-cluster-8-way-rugged-absolute-signature.html

11. Guidance Marine, '94-0561 SceneScan Guide', *Guidance Marine SceneScan User Guide.*

12. Guidance Marine, 'SceneScan Tracking and Performance', *Guidance Marine document 94-0574 – A.* http://www.guidance.eu.com/assets/_managed/cms/files/Brochure%202018/94-0574-A%20SceneScan%20Tracking%20Performance.pdf

13. D McKnight 'How targetless local position reference sensors will change and improve the experience of DP Operators', *Manuscript number SPE-192856-MS, ADIPEC* 2018

10. AUTHORS BIOGRAPHY

Sasha Heriot is Business Development Manager at Guidance Marine Ltd. After completing her PhD in Physics at the University of Sheffield in 2003, Sasha worked in the field of polymer physics as a research associate for 4 years. She left the world of academia to work in a spin out company from the University of Leeds specializing in novel surface treatments. She then worked for 3 years in a multinational chemical company as Market Development Manager for a specialist chemical business unit. Sasha started her appointment at Guidance Marine in April 2015 where she is currently introducing innovative technology into new markets and new applications.

David McKnight is Optical Technology Group Manager at Guidance Marine Ltd. Following a Master's degree in Physics at Oxford University David has worked in optical sensing in a variety of application fields. Prior to joining Guidance, he worked at Malvern Instruments designing nanoparticle characterisation systems and at PerkinElmer designing laser microscopy systems. In 2012 he joined Guidance Navigation to work on new technology for LIDAR based sensors.

Balint Furdek is Software Engineer at Guidance Marine Ltd. After completing his Bachelor's degree in Control Engineering at the Technical University of Cluj-Napoca, he worked in the Automotive industry at Continental Automotive (Romania) – developing SW for Nissan BCM modules. He joined Advanced Diagnostics (UK) – to work on key programming diagnostic tools. He joined Guidance Marine in early 2017 to assist with development on their latest sensors (CyScan AS, SceneScan).

Hendrik Busshoff holds the current position of Offshore Marine Superintendent at Bernhard Schulte Ship Management. As a trained Master mariner, he is responsible for offshore marine operations and DP. His previous experience includes SDPO on hyperbaric dive support vessel in the North Sea with Technip and sea time as officer and Marine Superintendent with the merchant navy.

AUTHOR INDEX